Leckie
the education publisher
for Scotland

T0187262

Higher
GEOGRAPHY

Revision + Practice
2 Books in 1

001/08012020

1 0 9 8 7 6 5

ISBN 9780008365301

Published by
Leckie
An imprint of HarperCollins*Publishers*
Westerhill Road, Bishopbriggs, Glasgow, G64 2QT
T: 0844 576 8126   F: 0844 576 8131
leckiescotland@harpercollins.co.uk   www.leckiescotland.co.uk

HarperCollins Publishers
Macken House, 39/40 Mayor Street Upper, Dublin 1, D01 C9W8, Ireland

Publisher: Sarah Mitchell
Project Managers: Janice McNeillie and Gillian Bowman

Special thanks to
Louise Robb (proofread)
Jouve (layout)
Sarah Duxbury (cover design)

Printed in Great Britain by Ashford Colour Press Ltd.

A CIP Catalogue record for this book is available from the British Library.

Acknowledgements

This product uses map data licensed from Ordnance Survey © Crown copyright and database rights (2014) Ordnance Survey (100018598). Pages 42, 50, 129, 134–35, 136, 169, 186–87.

The following were adapted from SQA questions with permission, Copyright © Scottish Qualifications Authority: Atmosphere question 1; Hydrosphere questions 1 and 2; Biosphere question 1; Population questions 1 and 2; Urban questions 1 and 2; Development and health questions 1 and 2; Global climate change question 2; Exam A Physical and Human Environments questions 5 and 11; Exam B Physical and Human Environments questions 2, 4, 3b and 5; Exam A Global Issues questions 1, 2 and 3.

References and text extracts: P101 The World Bank; P111 Met Office UK; P113 Climate Ready Clyde Review; P114 The Millennium Project; P117 Nature 2017; P118 The Environment Agency.

Illustrations © HarperCollins Publishers

Photos: P71 © epa european pressphoto agency b.v. / Alamy; P85 © lbert McCabe / Stringer / Getty Images; P89 Licensed under the Creative Commons Attribution 2.0 Generic license; P128 © Universal-ImagesGroup / Contributor / Getty Images; P163 1000 Words / Shutterstock.com. All other images © Shutterstock.com

Graph on P63 from National Records of Scotland; map on P120 licensed under the Creative Commons Attribution 2.0 Generic license.

Whilst every effort has been made to trace the copyright holders, in cases where this has been unsuccessful, or if any have inadvertently been overlooked, the Publishers would gladly receive any information enabling them to rectify any error or omission at the first opportunity.

MIX
Paper | Supporting responsible forestry
FSC™ C007454

This book contains FSC™ certified paper and other controlled sources to ensure responsible forest management.

For more information visit: www.harpercollins.co.uk/green

Contents

Introduction   2
PART 1: REVISION GUIDE
Section 1: Physical environments
Atmosphere   6
The global heat budget   6
The Earth's energy balance   8
Atmospheric circulation   10
Explaining the tricellular model   11
Oceanic circulation   12
Intertropical Convergence Zone (ITCZ)   14
Hydrosphere   17
Hydrological cycle (or water cycle)   17
River processes   21
Stages of a river   22
The upper course of a river   23
The middle course of a river   25
The lower course of a river   26
Hydrographs   27
The difference between describing and explaining a hydrograph   29
Lithosphere   31
Introduction to glaciation   31
Erosional features   33
Depositional features   35
Fluvio-glacial features formed by deposition   38
Glaciated landscapes on an Ordnance Survey (OS) map   39
Coasts   43
Coasts on OS maps   49
Biosphere   51
Soils   51
Soil formation   51
Factors determining soil type   52
Soil profiles and processes   53
Brown earth   54
Podzol   55
Gley   56

Section 2: Human environments
Population   58
The growing world population   58
Census – more than just a population count   58
Case study of a developing country: Nigeria   60
Other ways to collect demographic data   62
Population data presentation   63
Consequences of population structure   65
Migration   69
Causes and impact of voluntary migration   70
Causes and impacts of forced migration   71

Contents

Rural   **74**

Rural land degradation in North Africa   74

Physical causes of desertification   74

Human causes of desertification   75

Consequences of desertification   76

Possible solutions to desertfication   77

Case study: Tigray, Ethiopia   78

Case study: The Great Green Wall   78

Rural land use conflicts in a glaciated landscape: The Lake District   80

Urban   **83**

The urban population   83

Urban change and management in a developed world city: Glasgow   84

Changes to housing   85

Changes to transport in Glasgow   86

Urban change and management in a developing world city: Kibera, Nairobi, Kenya   88

Section 3: Global issues

River basin management   **91**

Global water surplus and deficiency   91

Why the world needs water management   94

How can river basins be managed?   95

Factors when considering the site of a dam   96

Multi-purpose river basin management   96

Case study: The Colorado basin, USA   97

Development and health   **101**

Indicators of development   101

Relationship between social and economic indicators   102

Inequalities between developing countries   104

Malaria: water-related diseases   105

Improving health care in developing countries   108

Example of WHO's work: primary health care (PHC)   108

Global climate change   **110**

Physical causes of climate change   110

Human causes of climate change   111

Local impact of climate change   113

Global impact of climate change   114

Management strategies and their limitations   116

Energy   **120**

The global distribution of energy resources   120

The reasons for changes in demand for energy in both developed and developing countries   122

The effectiveness of renewable approaches to meeting demands of energy and their suitability within different countries   123

The effectiveness of non-renewable approaches to meeting demands of energy and their suitability within different countries   125

Application of geographical skills   **128**

Glossary   **138**

Answers   **140**

Contents

PART 2: PRACTICE EXAM PAPERS

Revision advice   151

Practice exam A Physical and Human Environments   157

Practice exam A Global Issues and Geographical Skills   165

Practice exam B Physical and Human Environments   173

Practice exam B Global Issues and Geographical Skills   179

ANSWERS Check your answers to the practice test papers online:
www.collins.co.uk/pages/Scottish-curriculum-free-resources

ebook

To access the ebook version of this Revision Guide visit

www.collins.co.uk/ebooks

and follow the step-by-step instructions.

Higher Complete Revision and Practice

Complete Revision and Practice

This **two-in-one Complete Revision and Practice** book is designed to support you as students of Higher Geography. It can be used either in the classroom, for regular study and homework, or for exam revision. By combining **a revision guide and two full sets of practice exam papers**, this book includes everything you need to be fully familiar with the Higher Geography course and exam.

About the revision guide

The revision guide concisely covers the Higher Geography course content to help you prepare for both question papers in the final exam. There are 'Top Tips' throughout the revision guide which emphasise important points and give helpful hints. There is also an exam-style question in each content area that will allow you to apply your knowledge and understanding of the subject matter. Sample answers can be found on pages 140–148 to enable you to self-assess your response. There is also a glossary, which gives you a quick and easy way to find definitions of some of the key terms found in the revision guide.

How to use the practice exam papers

This book also contains two practice exam papers which replicate the layout, structure and question style of the Higher Geography exam as much as possible. There are also tips on important exam techniques that will help you gain valuable marks and avoid common mistakes. Sample answers for each of the questions can be accessed online at www.collins.co.uk/pages/Scottish-curriculum-free-resources. Responding to a range of questions is an effective study technique and we are confident you will find this book useful when preparing for the final exam.

 **Higher Complete Revision and Practice**

The Higher Geography course

The final course award you receive will be based on the marks you achieve in the two assessed components:

1. **Question papers**

    The question papers are worth a total of **160** marks:

    • Paper 1: Physical and Human Environments is worth **100 marks** and is scaled by SQA to represent 46% of the overall marks for the course assessment.

    • Paper 2: Global Issues and Geographical Skills is worth **60 marks** and is scaled by SQA to represent 27% of the overall marks for the course assessment.

2. **The assignment**

    The assignment is worth **30 marks** which represents 27% of the overall marks for the course assessment.

Higher

GEOGRAPHY

Revision Guide

Samantha Peck, Laura Sproule
Akiko Tomitaka

Atmosphere

In this section

- the global heat budget
- redistribution of energy by atmospheric and oceanic circulation
- cause, characteristics and impact of the Intertropical Convergence Zone (ITCZ)

Take a look at this website for more information:
http://www.georesource.co.uk/atmosphere.html

The Earth's atmosphere is made up of layers of gases that extend from the surface of the planet to the edge of space. Over 99% of our atmosphere is composed of nitrogen and oxygen, but other gases such as argon and carbon dioxide are also present. The Earth's atmosphere is responsible for our complex and dynamic weather phenomena and supports all life on the planet.

The global heat budget

The Earth's surface only absorbs 46% of the incoming solar energy (also known as insolation) that reaches the outer atmosphere. This energy loss can be explained by the conditions in the atmosphere **and** at the Earth's surface.

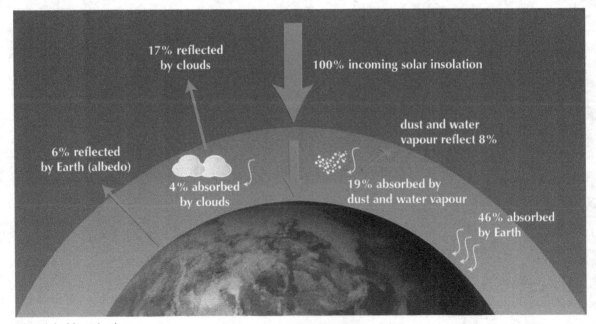

The global heat budget

Explaining the global heat budget

1. Conditions in the atmosphere

Clouds: Approximately 17% of the incoming solar insolation is **reflected** by clouds. Low, thick clouds reflect ~90% of solar energy whilst ~30% is reflected by lighter, higher clouds. More solar insolation penetrates through the atmosphere where there is less cloud cover. Only 4% of the incoming insolation is **absorbed** by clouds, as most of the energy is reflected.

Dust and gases: Dust particles and gas molecules reflect 8% of the initial solar insolation and **absorb** 19%.

This means that only 52% of the initial insolation reaches the Earth's surface.

2. Conditions at the Earth's surface

Roughly 6% of the initial solar energy is **reflected** by the Earth's surface due to the albedo effect. Light coloured surfaces reflect more insolation than darker areas and so the reflective potential can vary. For instance, the albedo can be less than 10% over oceans to ~85% over snow. Humans often modify the environment, which can alter the reflective potential of the Earth's surface.

The remaining insolation is absorbed by the Earth and is converted into heat energy. This amount is equivalent to 46% of the total energy received at the outer atmosphere.

Albedo refers to the reflective potential of a surface. The dark, dense vegetation of the rainforest absorbs more energy than the ice sheets of the polar regions.

The dark vegetation of the Amazon rainforest

The Earth's energy balance

Although on average the Earth absorbs only 46% of incoming solar energy, there are marked variations in this figure from the Equator to the poles. The graph below illustrates that areas within 37 degrees latitude of the Equator receive more solar insolation than is re-radiated. This creates a surplus of energy and is known as a **positive heat balance**. Areas beyond 37 degrees latitude of the Equator absorb less solar insolation than they emit energy. This creates an energy deficit and is called a **negative heat balance**.

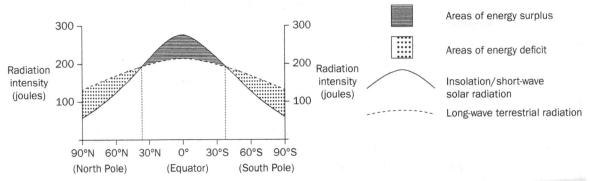

Energy balance and latitude

There are three main reasons to explain why there is an energy surplus in the tropical regions and a deficit in the polar regions.

1. The curvature of the Earth

2. The Earth's tilt and associated seasonality

3. The albedo effect

1. Curvature of the Earth

The diagram opposite illustrates that the Sun's rays are more dispersed at the poles due to the larger surface area (X), leading to less energy and colder temperatures. On the other hand, the rays are concentrated on a smaller surface area (Y) in the tropical regions due to the curvature of the Earth and the high angle of incoming solar insolation. This leads to an energy surplus near the Equator and warm temperatures.

There is also more atmosphere for the Sun's rays to travel through at the polar regions (A) than at the tropical latitudes (B) due to the low angle of the incoming insolation. This means that more energy may be potentially absorbed and reflected by the clouds, dust and gas molecules in the atmosphere. Ultimately, less insolation will reach the Earth's surface at the poles, contributing to an energy deficit and cold temperatures.

2. Tilt of the Earth and associated seasonality

At the tropical latitudes, the Sun is close to overhead throughout the year, and therefore energy is focused consistently on this area (see Graph A). This means that insolation levels are high in the tropical regions, leading to warmer temperatures. The Equator receives around 12 hours of daylight each day.

As the Earth is tilted on its axis, the poles receive much less daylight than the tropical regions for most of the year. For example, the North Pole is dark continuously for 6 months of the year and therefore receives no insolation during this time

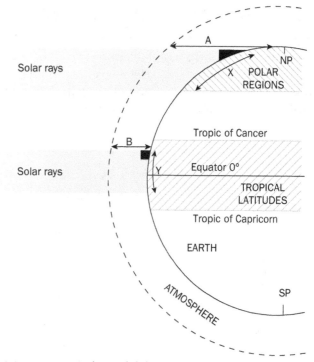

Explaining energy surplus and deficit

(see Graph B). This leads to lower average temperatures than at the tropical regions and contributes towards an energy deficit in the polar regions.

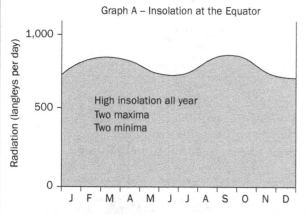

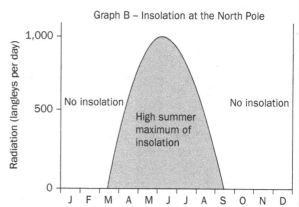

3. Albedo effect

In the tropical latitudes, more insolation is absorbed because of the high density of dark vegetation. On the other hand, more insolation is reflected at the poles due to the presence of light coloured, icy surfaces, which have a higher reflectivity potential (approximately 85% of solar energy is reflected).

EXAM QUESTION

With the aid of an annotated diagram or diagrams, **explain** why there is a surplus of solar energy in the tropical latitudes and a deficit of solar energy towards the poles.

10 marks

Atmospheric circulation

The global circulation model can be used to explain the Earth's transfer of heat energy from the tropical regions to the poles. As shown on the diagram below, the imbalance is addressed via cells: the Hadley, Ferrel and Polar cells. Without this crucial system, areas of energy surplus would simply overheat and our planet would look very different to how it is today.

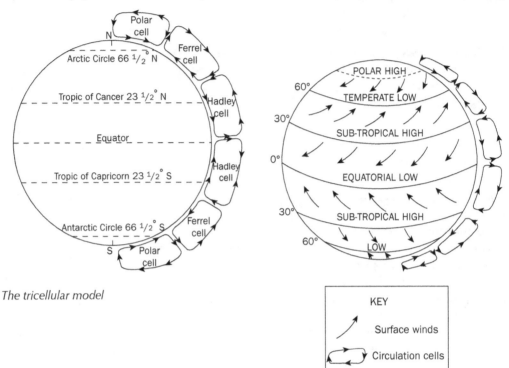

The tricellular model

KEY

↗ Surface winds

⟳ Circulation cells

Key idea:

The surface winds blow from areas of high to areas of low pressure. Analogy: When you blow up a balloon, you create a higher air pressure inside the balloon. When you let it go, the air inside needs to reach equilibrium with the air outside and wind is created.

TOP TIP

The concept of atmospheric circulation can be confusing. If you try to understand the reasons for high and low pressure systems, you will find the circulation model easier to understand.

Visit the Met Office's website to find out more:
https://qrgo.page.link/BS6mo

Explaining the tricellular model

Hadley cells

- On average, the equatorial regions receive the highest intensity of solar insolation. Here the surface air is warmer and less dense than the surrounding air.
- The warmer air has more kinetic energy, is less dense and therefore rises into the upper atmosphere (to roughly 18 km) creating an area of low pressure called the Equatorial Low. As the air rises and the water vapour cools and condenses, a band of clouds marking a distinction between the two Hadley cells is formed. This is known as the ITCZ.
- As the air reaches the upper atmosphere, it is forced to divert north and south of the Equator due to the up-draft from below.
- The air in the upper atmosphere then cools as it moves away from the site of most intense solar insolation. The cooler air is denser and sinks back to the Earth's surface at approximately 30 degrees north and south of the Equator. This forms a zone of high pressure called the Sub-Tropical Highs.
- Some of the air returns to the Equator as surface winds (also known as the trade winds) forming the Hadley cell.

Polar cells

- The movement of air in the Polar cells is driven by the Polar High.
- As colder air at the poles descends, it moves from high to low pressure across the surface to 60 degrees. This is known as the polar easterly winds. Here the air temperature rises, the air becomes less dense and rises, and the Mid-Latitude Low zone is created.
- When the rising air is deflected north, it completes the Polar cell.

Ferrel cells

- Unlike the other cells, the Ferrel cell is not driven by temperature. Rather, it acts like a cog in a machine, driven by the descending arm of the Hadley cell at approximately 30 degrees latitude and the rising arm of the Polar cell at approximately 60 degrees.

Low pressure

Rising air creates an unstable air mass, where water vapour condenses to form clouds and often precipitation. The low pressure zone around the Equator leads to high levels of rainfall, which explains the location of the world's tropical rainforests, such as the Amazon.

To learn more about the cells, visit **https://www.metoffice.gov.uk/weather/learn-about/weather/atmosphere/global-circulation-patterns**

High pressure

Sinking air creates a stable air mass, with limited or no cloud formation and a lack of precipitation. The Sub-Tropical Highs are crucial in determining the location of many of the world's deserts, such as the Sahara.

Oceanic circulation

It is not only our atmosphere that transfers heat energy around the world, but our oceans are also crucial in this redistribution process. But why do the oceans move in this manner?

- Ocean currents are greatly influenced by the prevailing (the most usual direction) winds. Energy is transferred by friction to the ocean currents.

- The Coriolis effect deflects currents to the right in the northern hemisphere and to the left in the southern hemisphere.

- As warmer water molecules expand and colder water molecules contract, the warmer water moves towards colder areas to address this imbalance.

- Cold water is denser than warm water, so the cold water sinks away from the poles towards warmer regions.

- There are salinity variations in the ocean (differences in the amount of salt in the water) which also has an impact on density, leading to the formation of currents.

- The currents are obstructed by continental land masses. The currents are deflected and can loop, forming gyres.

- The movement of seawater due to temperature differences and freshwater fluxes is also known as **thermohaline circulation**.

The humpback whale often takes advantage of the movement of ocean currents during its annual migration

Case study: Atlantic Ocean

- In the North Atlantic, a clockwise loop or gyre is formed when warm water from the Gulf of Mexico travels north-east. This is known as the Gulf Stream.

- The Coriolis effect deflects the currents clockwise in the northern hemisphere and anti-clockwise in the southern hemisphere.

- Prevailing southwesterlies drag the current further north-east. Colder water moves southwards, e.g. the Canaries Current.

- In the South Atlantic, water moves the opposite way than in the North Atlantic. Water moves southwards as the Brazilian Current and is deflected left by the Coriolis force.

- This movement of warm and cold water helps to maintain the energy balance.

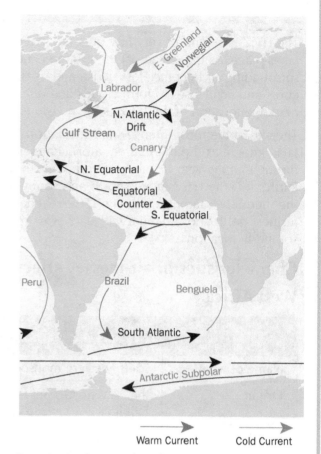

Oceanic circulation in the Atlantic

→ Warm Current   → Cold Current

TOP TIP

Try to remember key terminology and include it in your response:
- salinity
- gyre
- coriolis
- density

Perhaps you could write a glossary?

For more on the great ocean conveyor, visit
https://qrgo.page.link/GAUic

EXAM QUESTION

Explain the movement of ocean currents in the North Atlantic Ocean.

8 marks

Intertropical Convergence Zone (ITCZ)

The ITCZ is a band of very low pressure that forms where the most intense solar heating takes place. As the surface of the ground is heated by intense solar insolation, the surrounding air becomes less dense and rises. With increasing altitude, the water vapour cools and condenses to form clouds and often heavy precipitation. The band of clouds that mark the location of the ITCZ is often visible from space.

The site of the most intense solar insolation (also known as the thermal equator) changes throughout the year because of variations in the Earth's tilt. The ITCZ moves north and south with the thermal equator. In July, the ITCZ reaches its furthest north and in January, it reaches its most southerly location.

> ## TOP TIP
>
> You can get easy marks by recalling the characteristics of air masses. If you understand the reasons underpinning these characteristics, it will be much easier for you to remember them.

Characteristics of air masses affecting West Africa

	Tropical Maritime (mT)	Tropical Continental (cT)
Name of wind	south-western monsoon	harmattan
Origin	the Atlantic Ocean, in tropical latitudes	the Sahara Desert, in tropical latitudes
Weather	hot/warm temperatures, high humidity	hot/very hot, dry, low humidity
Nature	unstable	stable

Explaining rainfall patterns in West Africa

- In West Africa, rainfall patterns in the region vary considerably throughout the year. This is because the ITCZ moves seasonally, closely matched with the location of the thermal equator.
- Countries that lie south of the ITCZ will receive the wet weather brought by the Tropical Maritime air mass, while countries to the north of the ITCZ will be affected by dry weather brought by the Tropical Continental air mass.
- When the ITCZ is directly overhead, intense rainfall and thunderstorms are commonplace.
- When the ITCZ reaches its most southerly location across Africa, you may notice that it bends round the coastline. This is because the oceans absorb more heat than the land, thus the land area is hotter. This variation in temperature means that the ITCZ moves much further south while over land.

Thermal Equator:
This is the zone of the highest mean temperature over the Earth. Unlike the Equator, its position varies with the seasons.

Exam-style source material on the ITCZ

Below you will find maps and tables in the style often found in Higher Geography exam papers.

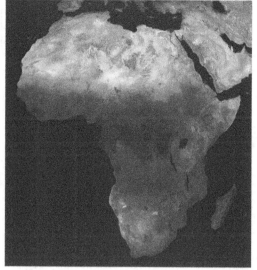

Satellite image of Africa. The presence of vegetation indicates areas of higher rainfall

The ITCZ

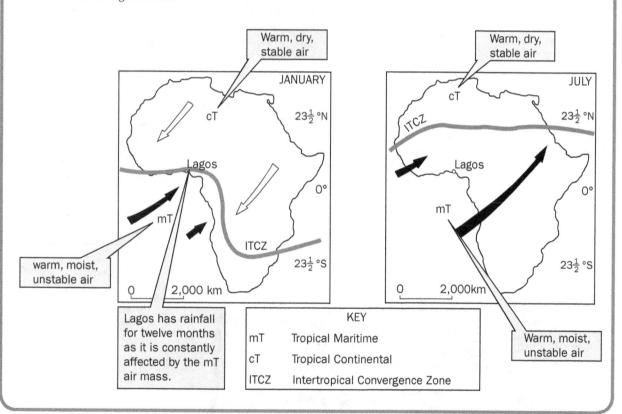

Warm, dry, stable air

JANUARY

cT

$23\frac{1}{2}$ °N

Lagos

0°

mT

ITCZ

$23\frac{1}{2}$ °S

warm, moist, unstable air

0   2,000 km

Warm, dry, stable air

JULY

cT

ITCZ

$23\frac{1}{2}$ °N

Lagos

0°

mT

$23\frac{1}{2}$ °S

0   2,000km

Warm, moist, unstable air

Lagos has rainfall for twelve months as it is constantly affected by the mT air mass.

KEY	
mT	Tropical Maritime
cT	Tropical Continental
ITCZ	Intertropical Convergence Zone

	J	F	M	A	M	J	J	A	S	O	N	D
Lagos	36	41	132	163	290	452	294	53	154	200	66	28
Jos	3	4	19	98	182	200	302	296	222	46	4	1
Timbuktu	0	0	3	0	3	25	80	82	50	3	0	0

Jos and Timbuktu have distinct dry seasons as they are affected by the cT air mass.

In Lagos, the two months with peak rainfall are June and October. This is due to the ITCZ migrating north and south with the thermal equator.

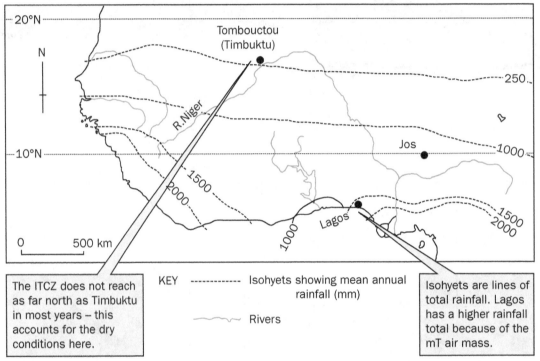

The ITCZ does not reach as far north as Timbuktu in most years – this accounts for the dry conditions here.

KEY ----------- Isohyets showing mean annual rainfall (mm)

~~~~~ Rivers

Isohyets are lines of total rainfall. Lagos has a higher rainfall total because of the mT air mass.

Typical ITCZ exam stimulus

TOP TIPS

- The ITCZ migrates north and south with the thermal equator. Don't write that it moves 'up' and 'down'.
- Try to refer to all of the sources provided in the question.
- Include data in your responses.
- Key terms include: thermal equator, air mass, mT, cT and peak rainfall.

GOT IT? ☐ ☐ ☐

Hydrosphere

In this section

- hydrological cycle within a drainage basin
- interpretation of hydrographs

Hydrological cycle (or water cycle)

There is a finite amount of water on the planet; this means that no water can be added or lost. However, water does constantly move between the oceans and the atmosphere. This means that the global hydrological cycle is a closed system.

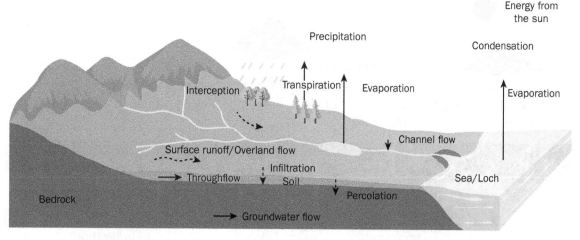

- Evaporation: moisture loss into the atmosphere from water surfaces, soil and vegetation.

- Transpiration: moisture loss through stomata (pores) in plant/tree leaves.

- Interception: raindrops are caught by vegetation before reaching the ground.

- Stemflow: raindrops then either drip off the leaves or flow down tree trunks or stems.

- Throughflow: The movement of water through soil towards the sea or river.

- Infiltration: the passage of water from the ground surface vertically into the soil layer.

- Goundwater: water stored in the pore spaces within a rock.

- Channel flow: the collection of water from rain directly falling in the channel, from surface runoff and from groundwater flow.

The hydrological cycle

The drainage basin

A drainage basin, on the other hand, is an open system, meaning it has inputs and outputs. A drainage basin is an area of land surrounding a river and its tributaries into which all the water drains. It will also include water that is stored in the water table and that flows over the surface as runoff. All rivers have an imaginary line, called a watershed, surrounding the land from which they receive water. When precipitation falls inside the watershed it will find its way into the river. If it falls outside of the watershed it will drain into a different river.

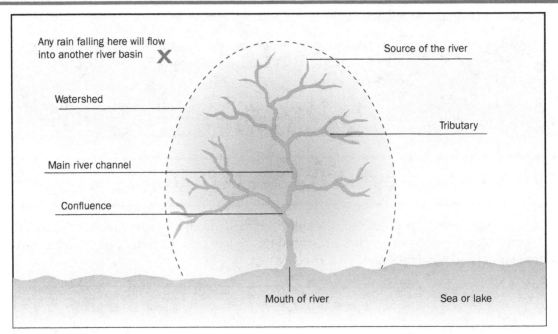

Any rain falling here will flow into another river basin **X**

Source of the river

Watershed

Tributary

Main river channel

Confluence

Mouth of river

Sea or lake

The drainage basin

The drainage basin hydrological cycle is an open system, where water is added and lost and constantly moves around. We say that the drainage basin hydrological cycle has inputs, outputs, stores and transfers.

Inputs	Outputs	Stores	Transfers
Precipitation	Evaporation	Rivers	Overland flow/ Surface runoff
Solar energy	Transpiration	Lakes	Infiltration
	Flowing into sea	Glaciers	Stem flow
		Soil	Throughflow
		Groundwater	Percolation
		Vegetation	

Physical factors affecting the hydrological cycle

The journey water takes as it transfers from one part of the hydrological cycle to another will be different in different locations around the world. Some of the physical factors causing these differences are shown in the diagram below:

Vegetation

Type – coniferous/deciduous woodland, grassland.

Cover – dense/intermittent/bare soil.

Relief

Steepness – this allows water to reach the channel faster.

Drainage basin

Shape – the more circular the basin, the shorter the lag time.

Size – the smaller the basin, the shorter the lag time.

Rock type (geology)

Permeability – permeable rocks allow water to pass through.

Porous rocks, e.g. sandstone, chalk contain pores that fill with and store water.

Pervious rocks, e.g. carboniferous limestone allow water to flow along bedding planes and joints even though they are impervious. There is less runoff as a result.

Impermeable rocks, e.g. granite are resistant to percolation and produce more streams.

Raindrop

Climate

Type of precipitation:

Prolonged rainfall – the ground becomes saturated. Infiltration is replaced by surface runoff.

Intense storms – heavy rain on hard surfaces or of an intensity greater than infiltration leads to surface runoff.

Snowfall – water held in storage. An increase in temperature leads to rapid melting and if the ground is frozen, infiltration is impeded.

Rate of evapotranspiration.

Determines the amount of water lost from the system.

Drainage density

Clays and impermeable rocks – high drainage density.

Sands and permeable rocks – low drainage density.

Soil type

This controls the speed of infiltration, the amount of storage and the rate of throughflow.

Sandy soils – less flooding.

Clay soils – more flooding.

Factors affecting the passage of a raindrop

Human factors resulting in interference to the hydrological cycle

There are a number of human activities that can interfere with the natural inputs, transfers and storage within a drainage basin:

- **Forestry**: increases interception.
- **Urbanisation**: the removal of natural vegetation to be replaced with impermeable surfaces such as concrete and drains can speed up overland flow and can lead to higher river levels. It also decreases the amount of water returning to groundwater storage, possibly reducing the water table.
- **Mining**: this might lead to a reduction in vegetation cover leading to increased runoff. Lakes, rivers and reservoirs may become silted up, leading to reduced storage capacity in these areas.
- **Deforestation**: the cutting down of trees increases runoff, decreases evapotranspiration and leads to extreme river flows as water is not intercepted and stored by trees.
- **Reservoirs/dam building**: interfering with the natural path of water can change the cycle quite dramatically. Building dams and reservoirs means less water is going underground and therefore stored underground. More surface water means increased evaporation and cloud forming, which may ultimately affect rainfall patterns.
- **Irrigation**: taking water from a river or underground store can reduce the river flow, lower water tables and increase evaporation by placing water in surface stores or by crops removing water from the cycle as they grow.

TOP TIP

Practice drawing the hydrological cycle and become familiar with all the labels and their meanings.

EXAM QUESTION

'A drainage basin is an open system with four elements – inputs, storage, transfers and outputs.'

Explain the movement of water within a drainage basin with reference to the four elements above.

10 marks

River processes

Rivers shape the landscape in three main ways: **erosion**, **transportation** and **deposition**. **Erosion** is when the river wears away land and the stones carried in it. **Transportation** is the movement of rocks and **silt** by the river. **Deposition** is when the river dumps rocks and silt wherever it slows down because it no longer has the energy to carry its **load**.

Erosion

A river erodes in four ways:

1. **Hydraulic action** – the breaking away of the river bed and banks by the sheer force of the water getting into small cracks and forcing pieces of rock to break off.

2. **Corrasion** – the wearing away of the river bed and banks by the river's load hitting against them and causing the landscape to break up.

3. **Solution/corrosion** – when the water in the river dissolves minerals from the rocks and washes them away.

4. **Attrition** – the wearing down of the load as the rocks hit the river bed and each other, breaking into smaller and more rounded pieces.

Transportation

Look at the figure below, which shows how a river carries its load.

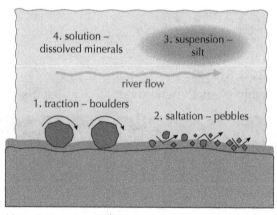

How rivers carry rocks

The river transports materials in four ways:

1. **Traction** – when large stones, e.g. boulders are rolled or dragged along the river bed by the force of the water.

2. **Saltation** – when smaller stones such as pebbles bounce off each other and are carried by the water.

3. **Suspension** – when small particles, e.g. silt are lifted in the water and carried long distances.

4. **Solution** – when the river dissolves minerals from the rocks that are carried in the water.

Stages of a river

A river has **three stages** called the **upper course**, **middle course** and **lower course**. It has specific features at each stage. Here we see the profile of a river and its valley.

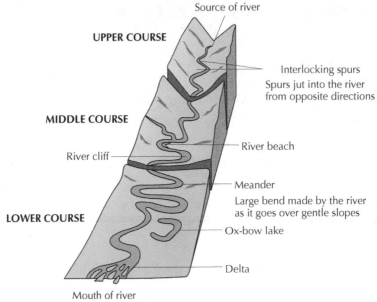

Model of the course of a river

Characteristics at each stage

Look at the table below, which lists the characteristics at each stage of the river.

Characteristics	Upper course	Middle course	Lower course
Slope	steep	quite steep	gentle
Width	narrow	quite wide	wide
Depth	shallow	quite deep	deep
Straightness	winding	meandering	large meanders
Amount of load	little	some	lots
Type of load	large/angular	medium and small/rounded	very small and rounded
Main work of the river	erosion and transportation	transportation	transportation and deposition
Valley width	narrow	quite wide	wide
Main features	V-shaped valley, waterfall, gorge	meander, river cliff, river beach/slip-off slope	flood plain, ox-bow lake, levées

The characteristics at each stage of a river

The upper course of a river

Landscape	Main process(es)	Main features
steep land	vertical erosion (downwards)	• V-shaped valley with interlocking spurs • waterfall • gorge

Characteristics of the upper course of a river

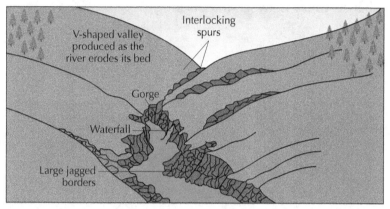

The upper course of a river valley

The formation of landscape features in the upper course of a river valley

V-shaped valley

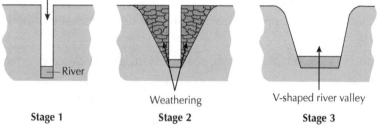

How a V-shaped valley is formed

- These are a characteristic of the upper course of a river valley.
- The greater altitude associated with the upper course of a river gives the river energy to vertically erode and cut down into its valley.
- The river cuts a deep gash into the landscape using hydraulic action, corrasion and corrosion.
- The shape of the V will depend on the rock type, climate and the type and amount of vegetation.
- As the river erodes downwards, the sides of the valley are exposed to weathering, e.g. freeze-thaw weathering. The rate of this weathering and other slope processes will

also determine the valley shape. Weathering will loosen rocks and they will fall into the river and over time be transported downstream. This helps to produce steep valley sides.

- Interlocking spurs form because the river is forced to follow a winding course around upland spurs of land.
- The rocks that have fallen into the river assist the process of corrasion, which leads to further erosion.
- The river transports the rocks downstream. The process of attrition helps to break rocks down and they become smaller and rounder.
- Over time, the valley becomes wider and deeper, creating a V-shaped valley between interlocking spurs.

Waterfall

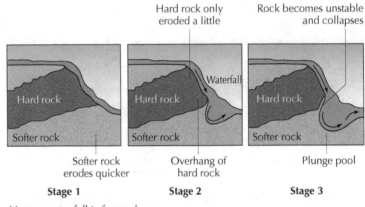

How a waterfall is formed

- These form in the upper and middle course where a river flowing over hard rock meets softer or less resistant rocks.
- Softer, less resistant rock is eroded away faster by the processes of hydraulic action, corrasion and corrosion.
- The river undercuts the harder rock, leaving it with an unstable overhang that may eventually collapse.
- The river erodes the softer rock below the waterfall by the process of hydraulic action. This is usually at times of high flow and a plunge pool is formed.
- The overhang of hard rock is unsupported and collapses into the plunge pool below.
- The process is repeated and the waterfall moves back upstream, leaving behind a gorge cut into the landscape.

The middle course of a river

Landscape	Main process(es)	Main features
moderate – gently sloping **gradient**	**lateral erosion** (sideways) and transportation	• meander • river cliff • river beach/slip-off slope

Characteristics of the middle course of a river

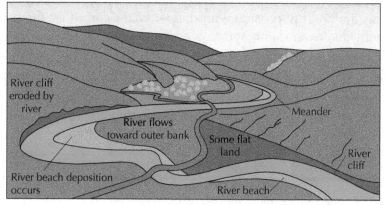

The middle course of a river valley

The formation of landscape features in the middle course of a river valley

Meander

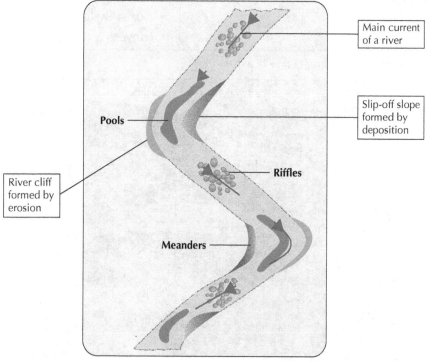

How a meander is formed

- A meander is a bend in a river where erosion takes place on the outside of the bend and deposition takes place on the inside of the bend.
- Meanders are a feature of the middle and lower course of a river.
- Water is naturally sinuous and so rivers rarely flow in a straight line. The main current of a river swings around areas of deposition or obstacles in the channel, e.g. a tree or hard rock.
- This results in areas of slower and faster water movement. Greater velocity (water speed) leads to erosion and slower velocity will lead to the deposition of transported material.
- Meanders are thought to be linked to the development of pools and riffles in the channel. Pools are areas of erosion with deeper water and riffles are areas of deposition with shallower depth and a steeper gradient.
- As the main river current flows from a pool to a riffle, it swings and this leads to it colliding and eroding the bank on the other side of the channel forming a river cliff.
- Helicoidal or spiral flow within the river cross-section moves material from the outside of the meander bend to the inside, reducing the depth and creating a slip-off slope.
- Over time, meanders become larger and more distinct.

The lower course of a river

Landscape	Main process(es)	Main features
flat land	deposition	<ul><li>flood plain</li><li>ox-bow lake</li><li>levées</li></ul>

Characteristics of the lower course of a river

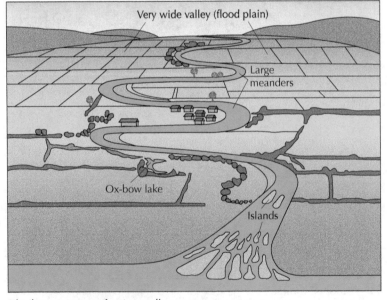

The lower course of a river valley

The formation of landscape features in the lower course of a river valley

Ox-bow lake

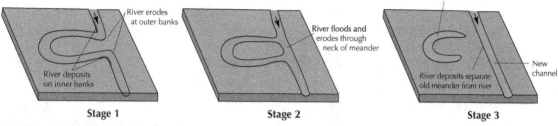

How an ox-bow lake is formed

- The river is **meandering** across the valley and **erodes laterally** (side of river bank).
- The river flows faster on the outside bends and erodes a **river cliff**.
- The river flows slowly on the inside bends and deposits material, forming a **river beach/slip-off slope**.
- Repeated erosion and deposition narrows the neck of the meander.
- During a **flood** or intense rainfall, the river will have more energy to erode and it **cuts through the neck of the meander**.
- The river flows on a new, straighter path and the meander is cut-off.
- The river **deposits silt** that seals off the ends of the meander and forms an **ox-bow lake**.

> **TOP TIP**
>
> Draw diagrams in pencil so you can easily correct mistakes.

Hydrographs

Hydrographs are a visual and graphical means of showing the discharge of a river at a given point over a short period of time.

It is important to understand how a river within a drainage basin will react to a period of rainfall as this can help predict whether a river will cope, or whether it will lead to flooding.

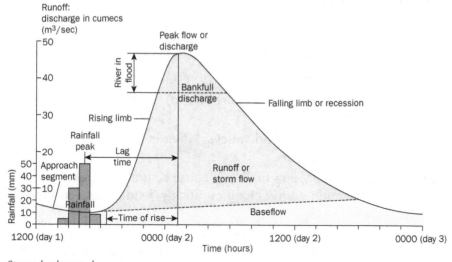

Storm hydrograph

The diagram shows a typical hydrograph. The line graph shows the discharge of a river (volume of water in the channel) and the bar graph shows the rainfall. Along the base of the graph is the time, which may be in hours or days.

Use this link to find out how you construct a hydrograph and what factors affect its shape. Remember to be careful about what scale you use.
http://www.georesource.co.uk/hydrosphere.html

When interpreting a hydrograph, you must mention some key terminology:

- **Peak flow/discharge**: occurs when the river reaches its highest level (shown by the highest point in the line graph).
- **Peak rainfall**: maximum rainfall (shown by the tallest bar).
- **Lag time**: the time between peak rainfall and peak discharge (it takes time for the rainwater to work its way over and under the ground to the gauging station where the discharge is measured).
- **Rising limb**: the steepness of the line graph indicates the speed at which the rainfall reaches the river. A very steep 'limb' is caused by rapid surface runoff reaching the river all at once.

TOP TIP

If you are asked to **describe** a hydrograph, you must make sure that you include all of these points:
- What time did is start raining?
- When was peak rainfall? How many millimeters?
- When did it stop raining?
- When did the discharge start to rise?
- What time was peak discharge? How many cumecs?
- What is the lag time?
- When did the discharge return to normal?

- **Falling limb**: the line graph as the river returns to normal. River discharge falls more gradually than it rises. Slower throughflow and groundwater flow feed the river more gradually and continuously than surface runoff.

Factors affecting the shape of a hydrograph

In some drainage basins, river discharge increases very quickly after a storm, which may lead to sudden, devastating flooding. In other drainage basins, the impact of a storm is absorbed better and a river's response is less extreme.

There are several factors that determine the level of response:

- **Basin size and shape**: the smaller the drainage basin, the more likely the rainfall will reach the river more quickly.
- **Relief and slope**: in steep-sided upland valleys, water is more likely to reach the river quickly than in gently sloping areas.
- **Types of precipitation**: prolonged rainfall will cause saturated ground, so rainwater will run over the surface and into the channel quickly, leading to a shorter lag time and higher peak discharge.
- **Rock type**: if the water falls onto an impermeable surface, the water will run over the land as surface runoff and will reach the river faster. Alternatively, if the rain falls onto a permeable surface, it will be infiltrated into the ground, which is a much slower process.

- **Land use**: dense vegetation cover will help by intercepting rainfall and this will lengthen lag times. Tropical rainforests intercept 80% of rainfall, arable land only 10%. Short lag times and high peak discharge are most likely to occur on bare ground surface such as deforested areas.

- **Temperature**: extremes of temperature can restrict infiltration, e.g. permafrost or baked desert soils.

- **Soil type**: clay soils will act as an impermeable barrier and rain will run off, whereas with sandy soils, infiltration will be greater, so lag times will be longer.

- **Urbanisation**: these areas tend to be covered in impermeable rock such as tarmac and concrete; therefore, the water will run over the land and reach the river quickly.

> ## TOP TIP
> If you are asked to **explain** a hydrograph, you must make sure you mention the factors above.

Basin lag time

The basin lag time (the time it takes for the rainfall to reach the river) depends on how fast or slow the processes are.

Fast processes	Slow processes
Overland flow	Infiltration
Surface runoff	Percolation
	Throughflow

The difference between describing and explaining a hydrograph

We can describe and explain this hydrograph as follows:

Description

- What time did it start raining?
  02.00 on 29 April.

- When was peak rainfall? How many millimetres?
  06.00 on 29 April, when 3.75 mm of rain fell.

- When did it stop raining?
  17.00 on 29 April.

- When did the discharge start to rise?
  03.00 on 29 April.

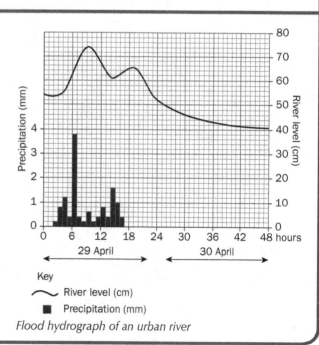

Flood hydrograph of an urban river

- What time was peak discharge? How many cumecs?

  09.00 when the river levels reached 74 cms.

- What is the lag time?

  3 hours.

- When did the discharge return to normal?

  River levels fall to 42 cm on 30 April.

Explanation

- The rising limb is quite steep and the lag time is short at only 3 hours, which shows that the rainfall reached the river very quickly. This could be due to it being an urbanised area; therefore, the rainfall will run over the land as surface runoff and into the river.
- As this is a town, there is likely to be little vegetation to intercept the rainfall.
- There will be little soil cover; therefore, rainfall will not be infiltrated.
- Plenty of impermeable surfaces such as concrete will increase runoff.
- The river basin may be small, allowing the rainfall to reach the river quickly.

EXAM QUESTION

Explain the changing river levels on the River Thaw at Cowbridge on 26 July 2007.

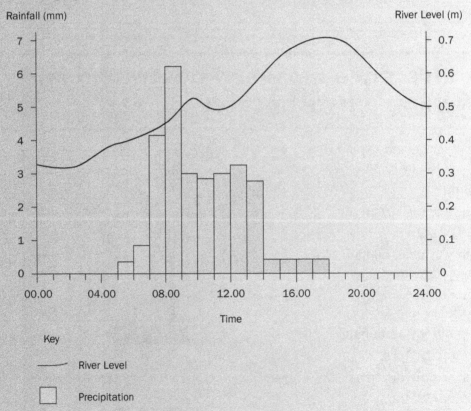

Key

— River Level

☐ Precipitation

12 marks

GOT IT? ☐ ☐ ☐
Lithosphere

In this section

- the formation of erosional and depositional features in glaciated landscapes
- the formation of erosional and depositional features in coastal landscapes
- Identifying features on an OS map

Visit **http://www.georesource.co.uk/glaciation1.html** for more photos and information on the glaciation topic.

Introduction to glaciation

The average temperature of the Earth has varied in cycles known as 'glacial' and 'inter-glacial' periods. During the last glacial period (which ended roughly 11,500 years ago), most of the landscape of the United Kingdom was covered in an icesheet. This ice shaped the landscape and many of the features left behind still exist today.

There are areas of the world that are still subject to glacial erosion and deposition, including the Alps, the Himalayas and the South Island of New Zealand.

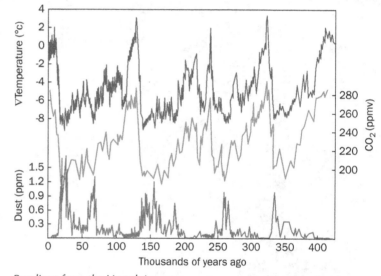

Readings from the Vostok ice core

Take a look at this collection of videos on the Ice Age and glaciers: **https://www.bbc.co.uk/programmes/b008g1kt/clips**

An active cirque glacier in New Zealand

Erosional features

Features of glacial erosion have been carved out by ice. You should be able to draw, label and explain how they have formed.

Time for Geography videos on glaciated landscapes:
https://timeforgeography.co.uk/videos_list/glaciation/

Formation of a corrie

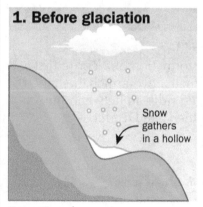

1. Before glaciation

Snow gathers in a hollow

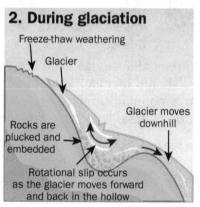

2. During glaciation

Freeze-thaw weathering

Glacier

Rocks are plucked and embedded

Glacier moves downhill

Rotational slip occurs as the glacier moves forward and back in the hollow

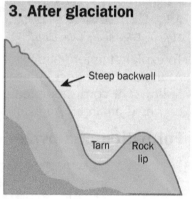

3. After glaciation

Steep backwall

Tarn   Rock lip

Formation of a corrie

- Snow collects in a hollow, which is often at a high altitude and north-facing as these areas are colder. The snow then **compresses, turning to firn/neve, then ice.**
- Due to gravity, this ice then moves downhill as a glacier.
- A Bergschrund crevasse may open up at the back of the hollow.
- **Freeze-thaw** weathering **erodes** the back of the hollow. This is when meltwater enters cracks in the rock, freezes and expands by approximately 9%. This forces the crack wider and when this has happened many times over, the rock will break apart.
- **Plucking** (when the glacier freezes onto the rock and plucks away the loose rock) occurs as it moves down the mountainside. This makes the backwall much steeper.
- The base of the hollow becomes deeper through **abrasion** (where pieces of rock, embedded in the bottom of the glacier, scrape and wear down the rock underneath as the glacier moves).
- After the ice melts, an armchair-shaped hollow is left in the mountainside; this is a **corrie**. A **lip** is found where less pressure is exerted by the glacier, or there has been

a temporary loss of energy and the glacier has had to deposit. Water can sometimes gather in the hollow and this is known as a **corrie lochan** or **tarn.**

- An example of a corrie is found on the **north-eastern side of Helvellyn**.

Formation of an arête

Striding Edge in the Lake District

Cross-sectional diagram of an arête

To explain the formation of an arête, you should write out the formation of a corrie first. At the end of your response you should include: after the ice melts, two armchair shaped hollows, or corries, are left in the mountainside; the knife-edged ridge in between the corries is an arête. An example of an arête is Striding Edge in the Lake District.

Formation of a pyramidal peak

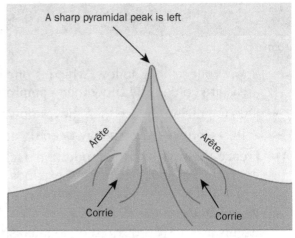

The Matterhorn

Formation of a pyramidal peak

To explain the formation of a pyramidal peak, you should write out the formation of a corrie first. At the end of your response you should include: after the ice melts, three or more armchair shaped hollows, or corries, are left in the mountainside; the highest point in the centre is called a pyramidal peak. An example of a pyramidal peak is the Matterhorn in Switzerland.

Formation of a U-shaped and hanging valley

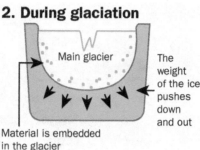

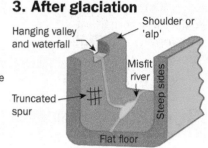

Formation of a U-shaped valley and a hanging valley

- Before glaciation, a river runs through a **V-shaped valley**.
- **Freeze-thaw** weathering weakens the rocks at the sides and base of the valley. This is when water enters the cracks in the rock. At night, when temperatures drop below zero degrees, the water freezes and expands (by as much as 9%). During the day when temperatures rise, the ice melts and releases the pressure. This is a continuous process that causes the rock to become very weak and break off.
- During glaciation, a glacier moves through the V-shaped valley. As it does so it pulls rocks away from the sides and base of the valley (known as **plucking**). **Abrasion** occurs when small rocks that are embedded inside the glacier scrape against the rocks, smoothing them like sandpaper. This deepens the base.
- After glaciation, a huge valley is left with very steep sidewalls and a very wide, flat base. The river will return but it no longer 'fits' the wide valley and so is known as a **'misfit stream'**. **Scree** often collects at the base of the valley sides due to weathering.
- **Hanging valleys** form from tributary rivers that flow into the main river below. During glaciation, these tributary valleys are filled with small glaciers. These glaciers do not have the same erosive power as the main valley glacier and so do less vertical erosion. After glaciation, the tributary valleys have been cut off by the main valley glacier and so are left 'hanging' above the main valley. When the rivers return, the hanging valley is usually marked with a waterfall.
- An example of a U-shaped valley is Great Langdale and a hanging valley is Little Langdale.

Depositional features

Depositional features are formed when the ice drops or deposits material.

Formation of terminal moraine
- Terminal moraine is a ridge running perpendicular to the valley.
- Moraine is the material that is carried by the glacier.
- Terminal moraines form when the ice loses energy and deposits all the moraine it was carrying. This might be when the glacier reaches flat land or when it begins to melt at low altitude where higher temperatures are found.
- The moraine is deposited at the front (snout) of the glacier.
- The longer that the ice continues to melt, the higher the terminal moraine.

- The presence of poorly sorted sediment (diamiction) is a characteristic of terminal moraine.
- Terminal moraines can extend for many kilometres.
- They mark the furthest point reached by the glacier (the maximum extent of glacial ice).
- As the glacier retreats, the moraine may form a dam.

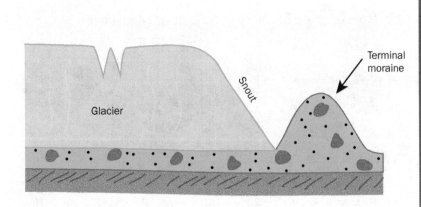

Formation of a terminal moraine

Formation of a drumlin

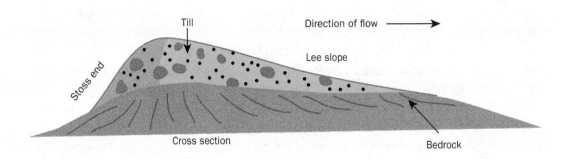

Cross-sectional diagram of a drumlin

- This is a feature of deposition made up of unsorted boulder clay/till (deposits from the ice sheet).
- The drumlin would have been deposited when the glacier became overloaded with sediment.
- The stoss end is steeper and refers to the side facing uphill.
- The lee slope is gently sloping and faces down the valley. This is because the glacier moves over an outcrop of bedrock and dumps material, before smoothing over the top.
- These are normally found in groups, known as swarms.
- They can be 1 km long and 500 metres wide.
- Glaciologists still disagree as to exactly how they were formed.
- Glasgow has been built on a series of drumlins.

Formation of a ribbon lake

- A ribbon lake can be a feature of erosion and/or deposition.
- A glacier forms in a hollow (often north-facing) where snow is compressed to form **firn/neve**.
- The glacier moves downhill due to weight and gravity.
- The glacier **plucks** rocks – freezes and pulls them away – from the valley sides, making them steeper.
- **Abrasion**, when the angular rock embedded in the ice grinds the bedrock, makes the valley deeper.
- Over time the valley becomes straightened, widened and deepened.
- Harder, more resistant rock at the base of the glacier is more difficult to erode, and softer, less resistant rock is therefore worn down by the processes of erosion. This creates a feature known as a rock basin.
- When this is filled with water, a feature known as a ribbon lake is formed.
- Terminal or recessional moraine deposited by the glacier may also dam the lake.
- An example of a ribbon lake is Lake Windermere.

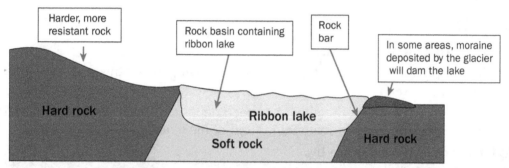

A cross-section of a ribbon lake

Lake Windermere in the Lake District

Fluvio-glacial features formed by deposition

Fluvio-glacial features are formed by sediment deposition from the glacial meltwater.

Formation of eskers

- Eskers are long, sinuous (winding) ridges of cobbles, gravels and sand that form across gently sloping landscapes.
- Eskers were molded in sub-glacial tunnels in which meltwater streams gradually accumulated bed deposits.
- On the wasting away of the ice, upstanding ridges were left behind.
- They run parallel to the direction of ice flow.
- Material is rounded due to the process of attrition.
- They vary in height and length.

A sub-glacial tunnel

Formation of outwash plain

- **Melting** ice produces vast quantities of water – flowing as streams from the snout of the glacier.
- These streams transport moraine that has been deposited by the ice and as the streams lose energy, they deposit the material in front of the glacier snout.
- As the deposits are laid down by water they are sorted. Finer material is carried further away from the margin of the ice.
- In addition, these outwash plains are often stratified, because the sediment is laid down in layers during annual flood events and during periods of higher discharge (in summer when there is more melting).
- Braided streams are often found in these outwash plains.

An outwash plain in New Zealand

Glaciated landscapes on an Ordnance Survey (OS) map

You may also be required to identify various features of glacial erosion on a map. The figures that follow show some examples of 3D landforms, with their associated contour pattern.

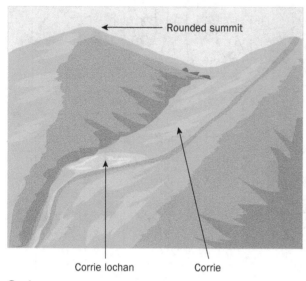

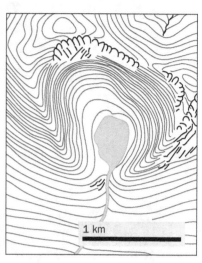

1 km

Corrie lochan   Corrie

Corrie

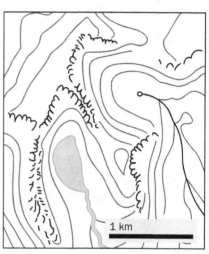

Arête

Corrie

1 km

Arête

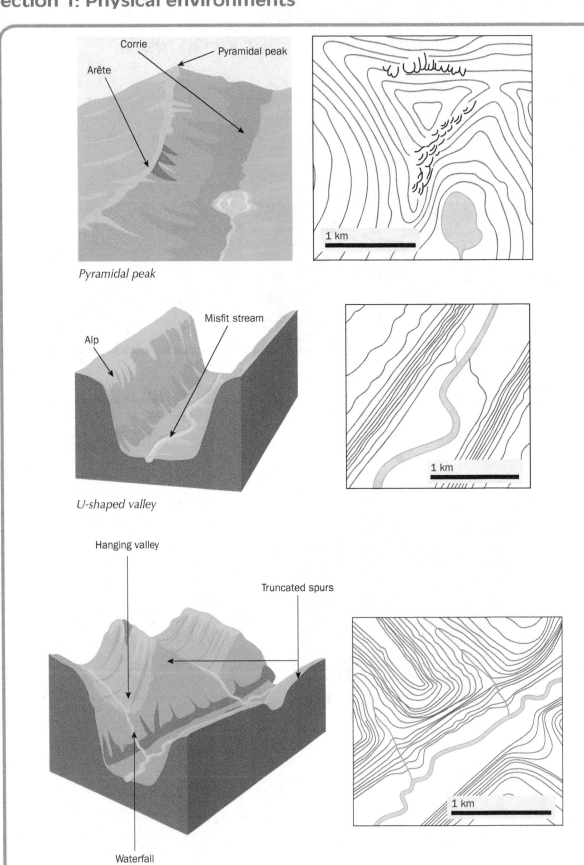

Corrie

Pyramidal peak

Arête

Pyramidal peak

1 km

Misfit stream

Alp

U-shaped valley

1 km

Hanging valley

Truncated spurs

Waterfall

Hanging valley

1 km

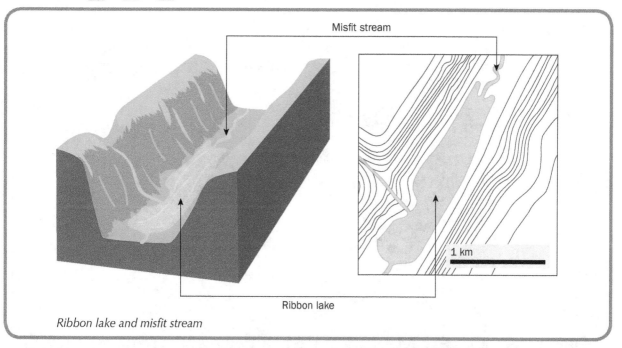

Misfit stream

Ribbon lake

1 km

Ribbon lake and misfit stream

An active glaciated landscape

EXAM QUESTION

Explain, with the aid of an annotated diagram or diagrams, how an arête is formed.

10 marks

The map below shows an area of the Cairngorm Mountains in Scotland. There are some hints below to help you identify glacial features on an Ordnance Survey map.

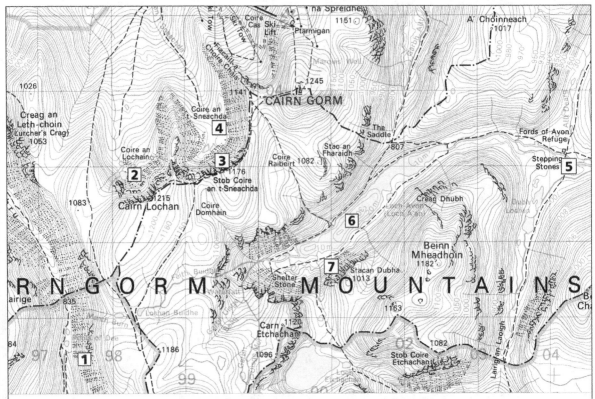

1. Scree: These are smaller rocks which are represented by black dots on the map. They are often found at the base of steep slopes.

2. Corrie Lochan: Coloured blue and found in a corrie – they are often named.

3. Arête: Look for a knife-edged ridge between two corries. Often surrounded by scree.

4. Corrie: These are often named so look out for the words Coire and Cwm. The contour lines are also horse-shoe shaped.

5. U-shaped Valley: Look for white space surrounded by contour lines which are very close together.

6. Ribbon Lake: Look for large bodies of water at the bottom of the valley.

7. Truncated Spur: This is where an interlocking spur is cut off. Look for exposed rock and scree at the side of the valley.

Features identified on an OS map

Other hints

U-shaped valleys often contain a misfit stream surrounded by flat land (represented as a white area on a map with few contours). Look out for roads as these are often built on the valley floor.

Pyramidal peaks are surrounded by three or more corries. Look out for a spot height at the top that signifies a peak, although not always a pyramidal one.

Truncated spurs can be difficult to spot. Look out for rock outcrops at the side of U-shaped valleys.

Hanging valleys are perpendicular to the main valley. Keep an eye out for waterfalls.

Coasts

Coastlines are the areas where the land meets the sea or the ocean. They are dynamic landscapes and are shaped by terrestrial (land) and marine processes.

Coastal erosion

There are many factors that determine the shape of a coastline, including: climate, rock type, tides, volcanic activity and wave frequency. Waves are the most effective agent of erosion along the coastline. They are created by wind blowing over the surface of the sea. Their size is determined by the fetch (the distance over which the wave has travelled), the wind speed and the depth of the water.

Constructive waves	Destructive waves
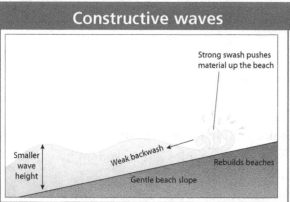	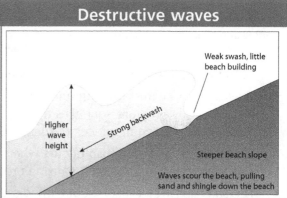

Constructive and destructive waves

There are four main erosional processes that you should be familiar with:

- **Hydraulic action**: waves crash against a cliff and drive water under great pressure into cracks in the rock. This pressure squeezes the air, and as the wave falls back, the air expands explosively, loosening pieces of rock.

- **Corrasion/abrasion**: occurs when stones and pebbles are picked up by waves and thrown against the cliff, causing erosion.

Time for Geography videos on coastal landscapes:
https://timeforgeography.co.uk/videos_list/coasts/

- **Attrition**: rock fragments hit against each other and so are reduced in size.

- **Solution**: salty sea water chemically dissolves rocks. This is most noticeable on chalk and limestone cliffs.

Formation of cliffs and wave-cut platforms

Photograph of wave-cut platform

- At high tide and during storms, waves erode cliffs at their base.
- Waves crash against the cliff and drive water under great pressure into cracks in the rock. This pressure squeezes the air, and as the wave falls back, the air expands explosively, loosening pieces of rock. This is called **hydraulic action**.
- **Corrasion or abrasion also** occurs when stones and pebbles are picked up by waves and thrown against the cliff, causing erosion.
- The salty sea water also chemically dissolves the base of the cliffs, in a process known as **solution**. This is most noticeable on chalk and limestone cliffs.
- These processes combine to form a wave-cut notch that eventually turns into a cave.
- The rock above the wave-cut notch is unsupported and cracks may begin to appear.
- Vertical cracks may eventually reach the top of the cliff, creating a blowhole.

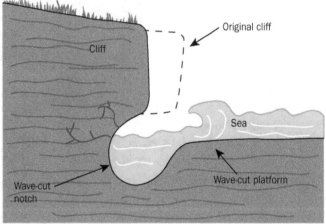

Diagram of wave-cut platform

- As the overhang is unstable, it collapses due to gravity. Weathering by wind and rain can also destabilise the cliff further.
- As the cliff retreats, a wave-cut platform is left. This is an area of flat land that juts out into the sea. It is often composed of more resistant rock.
- The whole process repeats and the coastline retreats further.
- An example is Durlston Cliffs.

Formation of headlands and bays

- Where there are different bands of rock, the coast is liable to differential erosion.
- Softer, less resistant rock is eroded more quickly through the processes of hydraulic action (the force of the waves against the rock), corrasion/abrasion (when stones and pebbles are picked up by waves and thrown against the cliff, causing erosion) and solution (when the salty sea water chemically dissolves the rock).
- The harder, more resistant rock is not eroded as quickly and 'sticks out' as a headland.
- Because the headland sticks out, it is subject to the highest energy waves. The bays though are sheltered and receive low energy waves that deposit rather than erode.
- This then means that the coastline becomes straight again and under the right conditions, the process may start again.
- An example is Swanage Bay and Peveril Point.

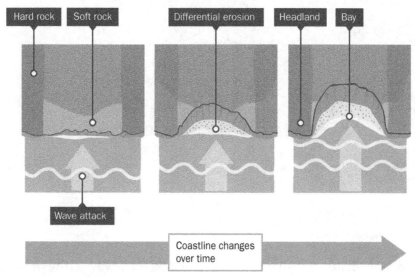

Formation of headlands and bays

Formation of caves, arches, stacks and stumps

Durdle Door in Dorset

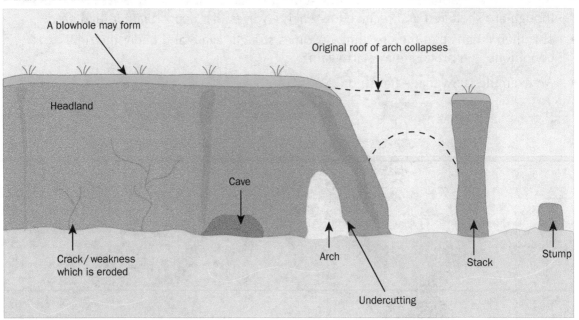

Formation of caves, arches and stacks

- When headlands made of resistant rock are attacked by waves, the processes of hydraulic action, corrasion/abrasion and solution occur.
- A cliff is undercut, forming a cave.
- When two caves form back-to-back, they meet to form an arch.
- The rock above the arch is now subject to hydraulic action by storm waves and to freeze-thaw weathering. It then becomes unstable and collapses due to gravity.

Longshore drift is explained in bite-sized form here: **http://www.bbc.co.uk/learningzone/clips/coastlines-longshore-drift/8440.html**

- This leaves a rock that is separated from the headland, called a stack.
- A stack will continue to be attacked by wave action and by physical and chemical weathering and is eventually worn down to sea level, leaving a stump.
- An example of a stack and stump is Old Harry's Rock and an example of an arch is Durdle Door.

Coastal transportation

When waves approach the coast, they do so at an angle because of the direction of the prevailing wind. Swash carries the material onto the beach, while backwash allows material to flow back to the sea at 90 degrees. As this process of transportation repeatedly occurs, material moves in a zigzag fashion along the coast. This movement of material is called longshore drift.

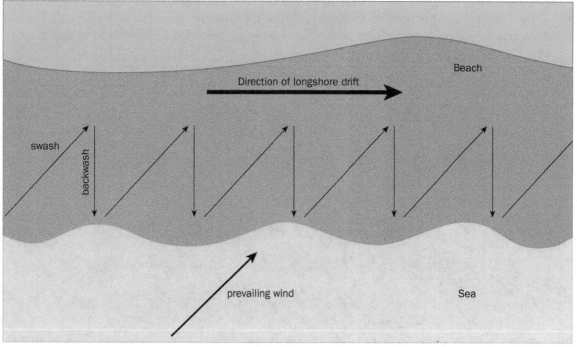

An aerial diagram showing the process of longshore drift

Coastal deposition

When the sea loses energy, it drops the material it has been carrying, depositing rocks, pebbles and sand. Coastal deposition is likely to occur in areas where waves enter a sheltered bay or shallow water, or when there is little wind.

Formation of sandspits, tombolos and bars

A tombolo

- Material is transported via the process of longshore **drift** along the coastline.
- Swash is where waves push material up the beach at an angle driven by the prevailing wind.
- The returning backwash is dragged back by gravity perpendicular to the beach.

- Where the coastline changes direction, the material will be deposited in open water.
- This material eventually builds up to above sea level and forms a sandspit (which is an extension to the beach).
- As the sandspit becomes longer, its rate of growth decreases because the water either becomes deeper, or the width of the channel is decreased and so the current becomes faster.
- If the sandspit connects with an island, it becomes a tombolo.
- If the sandspit connects with another piece of land across an inlet, it becomes a bar. A lagoon is often found behind a bar that will eventually fill with sediment and dry up.

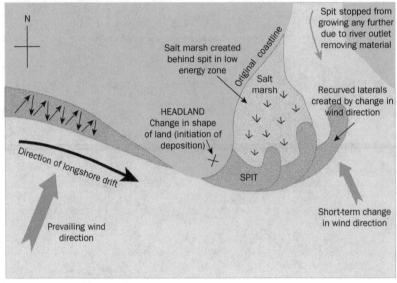

An aerial diagram of the formation of sandspits

Formation of a beach

- Beaches are found in areas where the deposition of sand and shingle is greater than its removal by wave action.
- They are often found in sheltered areas, such as bays, with low energy waves.
- Beach material is deposited by the swash as the waves lose energy.
- The backwash then returns the water to the sea, picking up some beach material.
- As the waves deposit material, the larger particles are dropped first.
- The beach therefore becomes graded, with the sand and silt often carried back into the sea.

Cape Tribulation in Australia

Coasts on OS maps

TOP TIP

You can determine the direction of longshore drift when sand spits are shown on a map. Material travels toward the end of the spit.

Just like in the Glaciation topic earlier in this chapter, you should be able to identify various coastal features on an OS Map. Here are some of the landforms you should be familiar with:

- Cliffs
- Wave-cut platforms
- Headlands
- Bays
- Caves
- Arches
- Stacks
- Sand spits
- Lagoons

For more on identifying these features, see **https://www.bbc.com/bitesize/guides/zsdmv9q**

The rugged coastline of the Isle of Skye

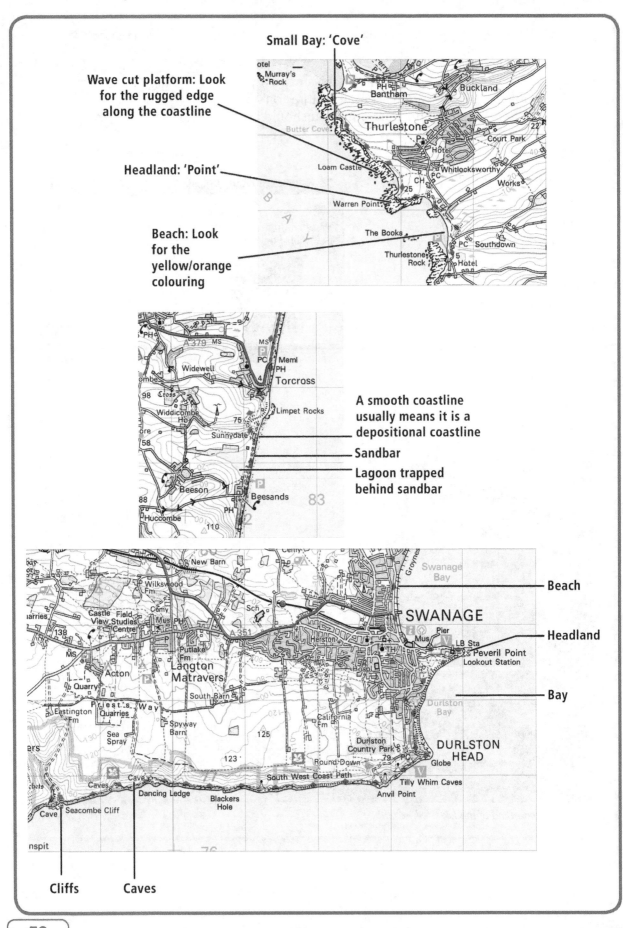

Small Bay: 'Cove'

Wave cut platform: Look for the rugged edge along the coastline

Headland: 'Point'

Beach: Look for the yellow/orange colouring

A smooth coastline usually means it is a depositional coastline

Sandbar

Lagoon trapped behind sandbar

Beach

Headland

Bay

Cliffs

Caves

Biosphere

In this section

- properties and formation processes of podzol, brown earth and gley soils

Soils

All forms of life on land depend one way or another on soil. Soil is the thin layer that has formed as a result of all the physical, chemical and biological weathering of the underlying rock (often called parent material) of the Earth's surface. It can vary in depth from a few centimetres to many metres. Yet its formation is slow; typically 1 cm of soil can take between 100–1,000 years to form, depending on the inputs and outputs.

For Higher you need to be able to describe the characteristics and explain the formation of three zonal soils (brown earths, podzols and gleys) all found in the UK, but also found in similar climates around the world.

You will need to know what the soil profile looks like for each of the soils. A soil profile is a vertical section through a soil (as if you had sliced a spade through the soil until you hit solid rock). Soil profiles are made up of soil horizons – these are distinctive horizontal layers in a soil profile.

Soil formation

The 'ingredients' of soil were identified by the American soil scientist H. Jenny in the following equation:

S= f(CORP)t

Soil is a **f**unction of **C** climate, **O** organisms and vegetation, **R** relief, **P** parent material and **t** time.

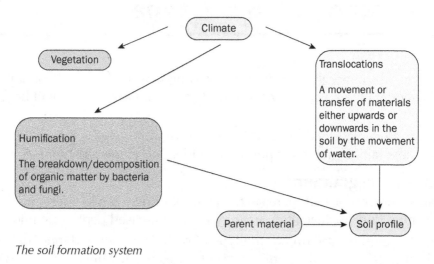

The soil formation system

Soils are made up as follows:

- **45% mineral matter**: this comes from the weathering or breakdown of the underlying parent material. It consists of a range of particle sizes from small clay particles (<0.002 mm diameter) to coarse sand (>2 mm diameter). The texture of the soil is determined by the particle mix.
- **5% organic matter**: this comes from decaying vegetable matter that is broken down by decomposers, e.g. fungi, earthworms.
- **25% air**
- **25% water**

Definitions of soil processes

Process	Definition
Acidity	A measure of the hydrogen ion concentration in the soil. Values greater than 7.0 are alkaline and values less than 7.0 are acid.
Capillary action	Transfer upwards of minerals through the soil horizons caused by evaporation loss at the surface.
Leaching	Downward washing by rainwater of soluble ions in solution.
Eluviation	The leaching of small suspended soil particles in infiltrating water from the A horizon.
Illuviation	The deposition of leached or eluvial particles into the B horizon.
Mor humus	Acid humus formed by the decaying of pine needles.
Mull humus	Soft, blackish organic matter formed by the decaying of deciduous leaves.

Factors determining soil type

Climate
- Temperature: determines the length of the growing season, the supply of organic material (amount of humus) and the speed of decomposition, which will be faster in warmer climates.
- Rainfall: where rainfall totals and intensity are high, there will be more leaching and in areas of less rainfall, more evaporation will lead to increased capillary action.

Organisms and vegetation
- Active micro-organisms will increase the amount of nitrogen fixation and the decomposition of dead vegetation, leading to an increased depth of humus.
- The type of vegetation determines the type of organic matter and therefore the pH. Most British soils are slightly acidic, particularly as heavy rainfall leaches out calcium, which is an alkali.

Relief

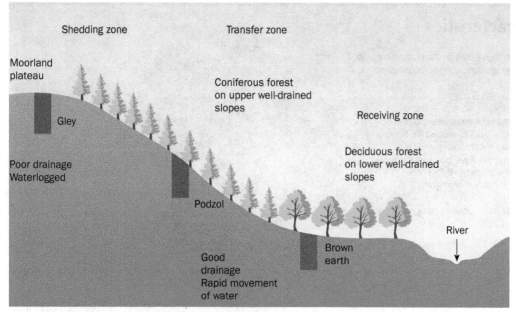

Diagram of a soil catena – this shows the relationship between soils and relief

Parent material

- This is the major factor determining soil type.
- It provides a supply of minerals and controls soil depth, texture, drainage (permeability) and soil quality.

Soil profiles and processes

For each soil type, you need to think about its **characteristics** (what it looks like), **formation** (key factors as to why it became, e.g. a brown earth and not a podzol) and **processes** (what is going on within the soil profile).

Brown earth

Characteristics

- Precipitation exceeds evapotranspiration.
- This soil is associated with deciduous forest.
- Generally pH 5 (slightly acidic).
- Pedalfer soil – so rich in hydrogen.
- Litter layer decomposes rapidly.
- A horizon is not bleached as rainfall totals are moderate.
- Clay minerals have been washed down and redeposited in B horizon.
- The high clay content gives it a distinctive texture.
- Good soils for cereal crops and good for grazing cattle.

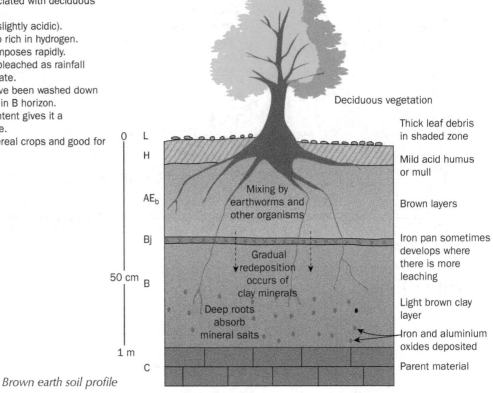

Brown earth soil profile

Formation

- Climate: temperate climates with moderate rainfall. Precipitation exceeds evapotranspiration.
- Vegetation: deciduous forests. Thick leaf litter layer that decomposes quickly.
- Organisms: active decomposers and worms, insects and rodents that help mix the soil.
- Relief: good drainage, possibly on a slope.
- Parent material: semi-permeable/permeable, good supply of minerals.

Processes

- Leaching, translocation.
- In the A horizon, clay minerals are washed down and redeposited in the B horizon.
- In areas of less rainfall the clay may remain, and only soluble minerals like calcium will move down.
- Iron or hard pans (these are thin layers of redeposited iron and form an impermeable barrier between horizons) sometimes develop where there is more leaching. Leaching is the removal of soluble minerals such as nitrogen (N), magnesium (Mg) and calcium (Ca) from the surface layers of the soil by the downward movement of acidic rainwater. Conditions that favour leaching are: heavy rainfall, free-draining soil and acid humus.
- Human use: good soils for growing cereal crops and good grazing land for cattle.

Podzol

Characteristics

- Occur where precipitation exceeds evapotranspiration.
- Coniferous forest/heathland vegetation in northern hemisphere.
- Usually colder climate so organic matter decomposes slowly.
- Pine needles form a thin litter layer so the humus is very acidic, which makes iron (Fe) and aluminium (Al) more soluble.
- Rainfall leaches the iron and aluminium from the A horizon leaving it a bleached ashy white.
- The iron is washed down into the Bf (iron pan) layer usually in winter.
- This hard, thin red or blackish iron pan layer is impermeable.
- This soil has limited agricultural potential but it can be improved by artificial drainage and the application of lime.
- Crops grown – oats, potatoes and hay.

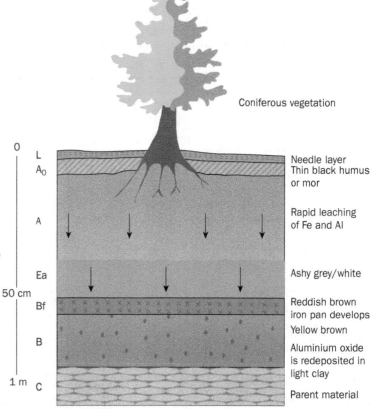

Podzol soil profile

Labels on profile:
- L
- A₀ — Needle layer / Thin black humus or mor
- A — Rapid leaching of Fe and Al
- Ea — Ashy grey/white
- Bf — Reddish brown iron pan develops
- B — Yellow brown / Aluminium oxide is redeposited in light clay
- C — Parent material

Depth scale: 0, 50 cm, 1 m

Coniferous vegetation

Formation

- Climate: high northern latitudes with high rainfall and cool temperatures. Precipitation exceeds evapotranspiration.
- Vegetation: taiga, coniferous forest and/or heathland. Thin litter layer of pine needles so the humus layer is very acidic, which makes iron (Fe) and aluminium (Al) more soluble.
- Organisms: cool temperatures mean decomposition happens slowly or not at all.
- Relief: free-drained upper slopes.
- Parent material: well-drained or sandy soils.

Processes

- Podzolisation, eluviations, illuviation, leaching.
- Podzolisation only takes place where you have an acid humus layer and intense leaching with acid rainwater passing through the A horizon and breaking down iron and aluminium oxide.
- Rainfall leaches iron and aluminium from the A horizon, leaving it a bleached ash-white colour.

- The iron is washed down (eluviated) in its blue (ferrous) conditon into (illuviated) the B horizon mainly in winter. As the soil dries in spring, oxygen can re-enter the soil and the iron oxidises to its red (ferric) state. The B horizon adopts an orange colour.
- The hard, thin red or blackish layer is iron rich and called the iron or hard pan. It is impermeable.
- Human use: these soils have limited agricultural potential but they can be improved by artificial drainage and the application of lime. Oats, potatoes and hay are the only cultivated crops on this soil type.

Gley

Characteristics

- Occurs in cold tundra areas with low precipitation and evapotranspiration.
- Generally shallow soils.
- Cold temperatures slow down decomposition so there is little organic matter.
- Gleying occurs when the output of water in the soil is restricted. The oxygen is quickly used up by micro-organisms and this deoxygenates the soil water and with the stagnant conditions the iron turns blue (ferrous).
- This leads to the soil being tinged blue or grey.
- However, mottled orange occurs whenever oxygen penetrates down the plant root channels or through cracks.
- A true gley is formed over a thick clay parent material.
- This type of soil can also form at the bottom of a slope or in a depression where drainage is poor.

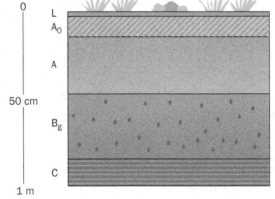

Gley soil profile

Formation

- Climate: temperate, tundra regions. Precipitation exceeds evapotranspiration. Cool to cold temperatures.
- Vegetation: grasses and low shrubs. Organic layer only partially decomposes.
- Organisms: few as too cold.
- Relief: poorly drained areas such as the bottom of a slope or in a depression. In tundra areas the permafrost means drainage is impeded.
- Parent material: usually thick clay and impermeable.

Processes

- Gleying.
- Oxygen depletion by bacteria in a stagnant waterlogged soil.
- This soil occurs when the output of water in the soil is restricted. The oxygen is quickly used up by micro-organisms and this deoxygenates the soil water and, together with the stagnant conditions, the iron turns blue (ferrous).
- This leads to the B horizon being tinged blue or grey.
- However, mottled orange occurs wherever oxygen has penetrated down a root channel or crack.
- Human use: these have very little economic potential.

This website has excellent photographs of the three soils along with details of their characteristics:
http://www.georesource.co.uk/biosphere.html

TOP TIP

Make sure you practice sketching and labelling all three soil profiles.

TOP TIP

Learn all the soil definitions carefully and be able to explain them in relation to all three soil types.

EXAM QUESTION

For a podzol, **explain** the main conditions and soil-forming processes that have led to its formation.

10 marks

GOT IT? ☐ ☐ ☐

Population

In this section

- methods and problems of data collection
- consequences of population structure
- causes and impacts of forced and voluntary migration

The growing world population

The world's population is growing fastest in the **developing** countries (e.g. Tanzania) and more slowly in the **developed** countries (e.g. United Kingdom). In countries like Japan, the population is decreasing.

According to the UN, the world's population will reach 9 billion by 2050 and 10 billion by 2100.

Population growth is the difference between birth rate and death rate. These figures are usually measured using per thousand. The symbol for per thousand is ‰.

7 Billion ...

The global population reached 7 billion in 2011

Census – more than just a population count

2011 Census

Take a look at this news article on the 2011 census:
http://www.bbc.co.uk/news/uk-12873011

The census is a survey carried out in the UK every 10 years by the government to collect information on:

- **Demographic data**: number of people, age and gender.
- **Social data**: marital status, citizenship, ethnic group, religion, languages spoken and education.
- **Economic data**: occupation, income, unemployment, type of housing, car ownership and mode of transport used to get to work.

Advantages and disadvantages of doing a census

Advantages	Disadvantages
The government can forecast demographic change and plan ahead on how to best spend and save the country's tax revenue (health care, social services, education, pension, housing, transport and other infrastructure).	It is very expensive – the last census in 2011 cost the UK government £480 million. The cost of carrying out a census is too expensive for some developing countries.
Academics use the data to study the changes in society.	It takes at least two years to process the information – by that time the information is out of date.
Genealogists use old censuses to trace ancestry.	Result isn't 100% accurate – some people write false information or do not return their census at all.

Reasons for inaccuracies in developing countries

- **Size of the country**: some countries' territories are vast – for example, India is 13 times bigger than the UK, which makes it very difficult to cover every part of the country.
- **The terrain**: mountainous countries like Nepal and Pakistan have poor or no roads to access the population in remote areas.
- **The climate**: some roads become impassable during hurricane/rainy season.
- **Nomadic population**: around 4% of Tanzania's population is considered to be nomadic (Maasai and Luo). Nomadic people move frequently as part of their traditional lifestyle, which makes it very difficult to get an accurate count of the population.
- **Literacy rate**: Afghanistan's literacy rate is only 32%, which means the vast majority of the population is illiterate and therefore can't read the census questions and can't provide a written answer. The government will have to spend money on employing enumerators to conduct the census orally. People are less willing to give private information to enumerators, which leads to more inaccuracies.
- **Communication**: in some countries there are many languages spoken – India for example has more than 20 official languages, and in Papua New Guinea 820 different indigenous languages are spoken. Censuses will have to be translated, which adds to the cost, and some questions may not be translated well or accurately.
- **Lack of trust**: some countries have suffered major civil unrest (like Rwanda in 1994). This could mean people are less likely to trust the government and therefore less likely to answer the questions honestly.
- **Cost of carrying out the census**: most developing countries lack money and other resources to do a census every 10 years. They also lack the technology to process the information efficiently.
- **Inflated numbers**: community leaders may encourage the local population to write inflated population numbers in the hope of getting more funding for services from the central government.

- **Exceptional events**: in the summer of 2014, much of West Africa was under threat from the Ebola virus. Sierra Leone was scheduled to hold a census in December 2014 but this has now been postponed until at least December 2015. While epidemics can happen anywhere in the world, developing countries find it harder to control the diseases due to lack of resources.
- **Current civil unrest**: Syria was scheduled to hold a census in 2014, but due to the continuing civil war it currently has no plans to conduct a census.

Case study of a developing country: Nigeria

Nigeria is often referred to as the 'Giant of Africa' as it is the most populous country on the continent and the 7th most populous in the world. The Nigerian government find it challenging to collect accurate population data because of the following complex factors:

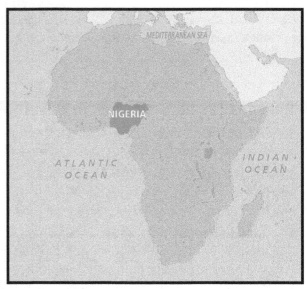

Nigeria in the continent of Africa

- Nigeria is twice the size of the UK. This makes it difficult to reach everyone in the country.

- Nigeria's roads and railways are poorly maintained. Some rural settlements are very difficult to reach, especially in the rainy season when roads become impassable.

- Nigeria's border with Cameroon has a sizeable mountain range called the Atlantika Mountains, which is remote and not easily accessible.

- Nigeria is home to over 18 million nomadic tribespeople, known as Fulani (who make up around 10% of the population). They make a living from herding cattle, goats and sheep. The majority of the Fulani migrate around West Africa in search of grazing ground for their animals. North of Nigeria is also home to another nomadic tribe called the Tuareg. Nomadic people usually do not have a fixed settlement and they move around West Africa throughout the year, making them difficult to count.

For more information on Makoko, have a look at this report: **https://www.france24.com/en/20130830-reporters-nigeria-shanty-town-lagoon-makoko-lagos-fishing-community-houses-demolition**

- Lagos is Nigeria's largest city. There are approximately 100,000 people in Lagos who live in a shanty town called Makoko. The shanty town population is growing rapidly due to rural–urban migration. To count people in the shanty town is a difficult task because the residents do not have a proper address to which the census form can be sent, or an enumerator can visit. Due to the recent threat of eviction from the mayor of Lagos, people in the shanty town distrust the local government and are therefore less willing to answer questions.

- Although the 'official' language of Nigeria is English, there are 500 other languages and dialects spoken in Nigeria. This makes it difficult and costly to translate the census questions. Due to limitations with literal translation, certain words can be mistranslated between languages and this can lead to inaccuracy.

- Only 61% of Nigerians are literate. This means the government will have to employ larger numbers of enumerators to go around each settlement to ask questions in person. This is time consuming and expensive.

- Even in a modern democratic country, Nigeria's politics is divided down tribal lines. People from a certain village or tribe may give inflated numbers for their population in order to obtain more resources from the central government or to increase political representation.

- Three northern states in Nigeria have been attacked by a terrorist group called 'Boko Haram'. The civil unrest has led to displacement of 250,000 people and a further 3 million people are facing a humanitarian crisis. In such a politically unstable environment it will be very difficult to obtain accurate data as people may feel frightened to answer ethnic or religious questions one way or another.

- Even once the census data is collected, developing countries like Nigeria lack the resources and statistical experts to process such a large volume of data into useable information, which defeats the purpose of carrying out a census in the first place.

- Nigeria's last census in 2006 is largely seen as an inaccurate count by population experts.

You can read more about the Nigerian census here: **http://news.bbc.co.uk/1/hi/world/africa/4512240.stm**

Other ways to collect demographic data

Civil register

Most countries have a civil register: birth, marriage and death certificates (also known as **vital statistics**) are legal documents. In Scotland, births, marriages and deaths need to be registered within 21 days of the event.

Government departments

The UK and Scottish government statistical departments are:

- Office for National Statistics and UK Statistics Authority are responsible for collecting and publishing statistics related to the economy, population and society at national, regional and local levels.

- Scottish Neighbourhood Statistics is the Scottish Government's on-going programme to improve the availability, consistency and accessibility of statistics in Scotland.

- Since 2004, the UK government has collected additional population data by conducting an annual population survey. This is a survey carried out on a small random sample of people (360,000 individuals). Its purpose is to provide information on key social and socio-economic variables between the 10-yearly censuses, with particular emphasis on providing information relating to sub-regional (local authority) areas.

Organised research

There are also research organisations that deal with population statistics:

- Market research companies like Ipsos MORI, YouGov and ICM collect data and people's opinions about current affairs. Public organisations and private companies often use the findings of of these companies to make decisions about goods and services.

- The Institute of Public Policy Research (IPPR)'s purpose is to conduct and publish research into, and promote public education in the economic, social and political sciences, and in science and technology; including the effect of moral, social, political and scientific factors on public policy and on the living standards of all sections of the community.

Loyalty cards

Information about our shopping habits can be used to compile economic data.

Immigration

- Police registration is required for some nationalities as part of their visa condition.

European and global bodies

Eurostat collects European Union-wide data on economics, population, trade, industry and agriculture. The UN Statistical Division collects world-wide data on social, economic and environmental issues.

EXAM QUESTION

Explain the problems of collecting accurate population data in developing countries.

12 marks

Population data presentation

Two of the most useful ways to show population structure and change are the demographic transition model (line graph) and the population pyramid (bar graph).

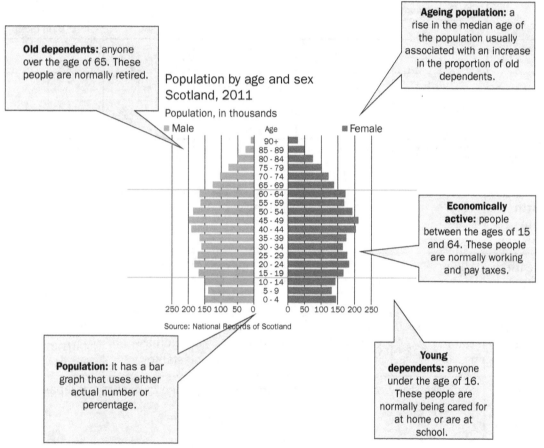

Old dependents: anyone over the age of 65. These people are normally retired.

Ageing population: a rise in the median age of the population usually associated with an increase in the proportion of old dependents.

Population by age and sex
Scotland, 2011

Population, in thousands

■ Male   Age   ■ Female

Economically active: people between the ages of 15 and 64. These people are normally working and pay taxes.

Source: National Records of Scotland

Population: it has a bar graph that uses either actual number or percentage.

Young dependents: anyone under the age of 16. These people are normally being cared for at home or are at school.

Population pyramid. If you want to see population pyramids of other countries, visit: www.populationpyramid.net

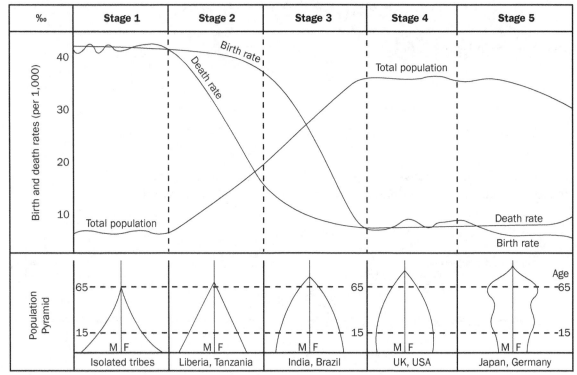

Demographic transition model

	Stage 1	Stage 2	Stage 3	Stage 4	Stage 5
Birth rate (BR)	High	High	Falling	Low	Very low
Death rate (DR)	High	Falling rapidly	Continues to fall	Low	Low
Natural increase	Very little / stable	Rapid / very rapid increase	Continues to increase at slower rate	Slow increase / stable	Slow Starts to decrease
Reasons for changes in BR	See case study of Tanzania		Increased quality of life and standard of living	See case study of Japan	
Reasons for changes in DR	Diseases, famine. Poor medical knowledge and access, resulting in high infant mortality and low life expectancy.	Improvement in medical care and access to care. Cleaner water supply and better provision of sanitation.		Good health care. Good standard of living and quality of life.	

Consequences of population structure

Case study of an ageing population: Japan

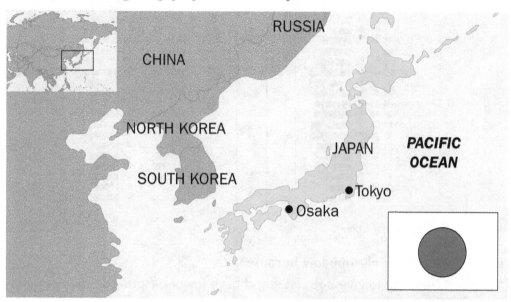

Japan is at stage 5 of the demographic transition model (DTM) and it is the fastest ageing country in the world; the baby boomer generation (people born 1947–49) have reached retirement age.

Japan's birth rate is lower than its death rate – this means the population is ageing and declining. For the last 30 years, the number of babies born in Japan has been decreasing.

You can read more about Japan's falling birth rate here: **http://www.bbc.co.uk/news/world-asia-30653825**

Japan: Population data	
Population	126.5 million
Birth rate	7.6‰
Death rate	10.8‰
Natural increase	–0.2%
Life expectancy	84 years
% of population age 15 and under	13%
% of population age 65 and over	27%
Fertility rate per woman	1.4
% of urban population	92%
GNI (gross national income) per capita	US$45,000
DTM stage	Stage 5

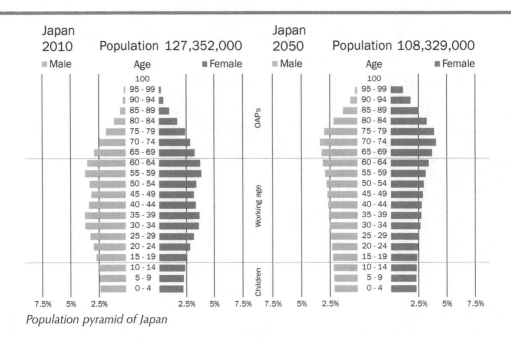

Population pyramid of Japan

Japan's birth rate is at an all-time low because:

- Women are choosing to give birth later and have fewer offspring so they can pursue their ambitions, e.g. for further education and having a career.

- In most developing countries, children are seen as economic burdens as it is very expensive to raise a family. Japan is one of the most expensive places to buy a house, and most people cannot afford to buy their own house so they rent throughout their lives.

- Although education is free, it is very competitive to get into prestigious universities so parents have to spend a lot of money on extra tuition and extra-curricular activities.

- Japan's maternity leave is only 14 weeks and during this time the new mothers are paid 67% of their normal pay. Job security for new mothers is not 100% and they are less likely to be promoted later in their careers.

Some consequences of the low birth rate are:

- The population of Japan is decreasing. It is estimated that in the next 50 years the country's population will shrink by a third.

- Fewer babies being born means less money needs to be spent on pre- and post-natal care. Some maternity hospitals will close and some midwives will lose their jobs.

- Fewer children also means less money needs to be spent on childcare and education. This will lead to some schools amalgamating or closing and some teachers will lose their jobs.

- Redundant doctors, nurses and teachers will have to be paid unemployment benefits and they may also need to be retrained for other professions. This is very expensive for the government.

- In the long term, companies will find it difficult to recruit young workers.

- The profits of companies producing goods and services for babies and children may suffer.

Due to a healthy diet, lifestyle and good health service, Japan's life expectancy is one of the highest in the world. In 2013, Japanese people aged 65 or older were at a record high, making up a quarter of the population.

Consequences of an ageing population are:

- A low fertility rate means a lower number of economically-active members of the population. This means fewer working-age people will have to support a larger ageing population, meaning the government will have to raise taxes.

- Old people are more likely to be ill and require medical care. More resources will need to be spent on geriatric services. Old people also require more assistance with day-to-day tasks, so more nursing homes and day-care centres will have to be built.

- A smaller economically-active population means Japan's economy will suffer from a lack of skilled workers, making the country less competitive in the global market.

- Until recently, the mandatory retirement age in Japan was 60 but the government has raised this to 65.

- Older people are encouraged to stay on working after their official retirement age at reduced hours and responsibility. There has been a significant increase in the 'silver workforce' in Japan. Older workers are valued for their maturity and experience.

- Since April 2019, the Japanese government have relaxed the working visa application for both skilled and unskilled workers from abroad.

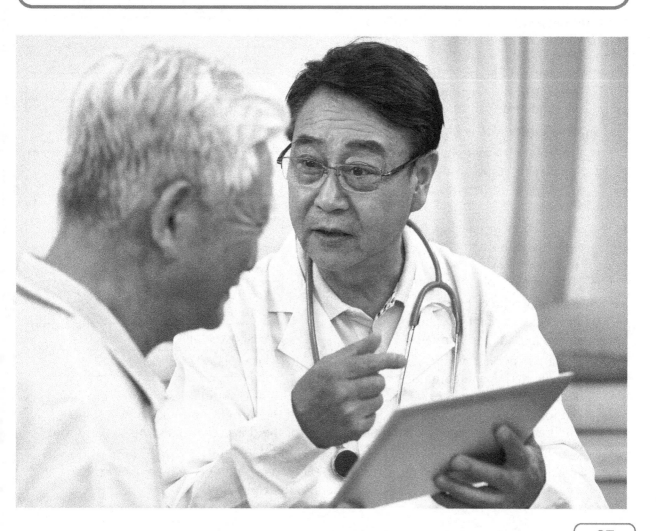

Case study of a rapidly growing population: Tanzania

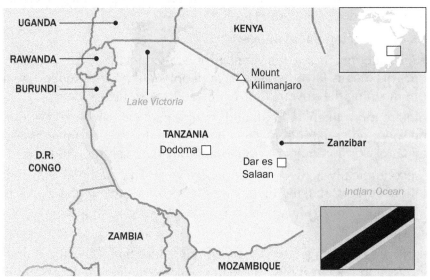

Map showing the location of Tanzania

Tanzania is a developing country located in East Africa. It has a high birth rate and a low death rate, which means the population is growing rapidly: it is at the end of stage 2 of the DTM. Tanzania has a young population with an average age of 17. This means that the majority of the population is yet to get married and have babies.

Tanzania's birth rate is high because:

- Tanzania is a traditional country where many people still hold conservative values. In such societies, large families are considered to be normal.

- Many people in Tanzania hold conservative religious values and it is largely considered to be taboo to have an abortion. Approximately 10 million people in Tanzania are Catholics.

- In developing countries it is more difficult and costly to obtain contraception and advice on family planning.

- Tanzania's economy is heavily based on agriculture. In rural communities 85% of the population is engaged in farming. In traditional rural communities children are seen as economic assets, as they can work and help on the farm from a relatively young age. Children in Tanzania also take on their share of the domestic duties and some go out to work from a young age to supplement the family income.

- Tanzania doesn't have a state pension for everyone so parents will have more children to make sure someone will look after them in their old age.

Tanzania: Population data	
Population	56.3 million
Birth rate	37.7‰
Death rate	6.5‰
Natural increase	3%
Life expectancy	66 years
% of population age 15 and under	45%
% of population age 65 and over	3%
Fertility rate per woman	5
% of urban population	34%
GNI per capita	US$3,160
DTM stage	End of stage 2

- The female literacy rate in Tanzania is 70%, which means one-third of the female population cannot read and write. Illiterate girls go on to marry early and produce a large number of children. On average a Tanzanian woman will give birth to 4–5 children in her lifetime.

- Due to poverty and lack of education, infectious diseases such as malaria and cholera kill a lot of newborns. The infant mortality rate in Tanzania is high (38‰). Parents have lots of children to compensate for the high infant mortality rate.

The consequences of a rapidly growing population are:

- Overcrowding in urbanised areas due to the large number of young people in search of a job. This leads to the growth of shanty towns and informal settlements.

- Unemployment or underemployment may be high due to a large workforce and not enough jobs.

- The government needs to build more schools and hospitals. The Tanzanian government do not have the money to build enough facilities, so people suffer from poor services.

- More teachers need to be trained in order for everyone to obtain a decent quality of education.

- The government must work hard to attract investment from national and international companies to create more jobs. The government will continue to receive foreign aid to fund the programmes needed to develop the country.

Migration

Reasons for moving away from a person's original location are called **push factors**, and reasons to move to another location are called **pull factors**.

Common push and pull factors

Push factors	Pull factors
Unemployment	Potential for employment
Lack of safety	A safer atmosphere
Lack of services	Better service provision
Poverty	Greater wealth
Crop failure	Fertile land
Drought	Good food supplies
War, civil unrest	Politicial security
Hazards	Less risk of natural hazards
Isolation	Friends and family

Migration can happen within the same country; this is known as **internal migration**. Migration can also be between two different countries and this is known as **international migration**.

Causes and impact of voluntary migration

Case study: Poland to Scotland

Characteristics

- Poland and Scotland have close historical ties.
- Today, people born in Poland constitute the largest non-UK-born group in Scotland. It is estimated that 56,000 residents of Scotland were born in Poland.
- In 2004, Poland joined the European Union and since then the number of Polish-born residents in Scotland has increased due to ease of movement and right to employment.

Migration from Poland to Scotland

Incentives to move

Push factors:

- Lack of jobs – in 2004 the average unemployment rate in Poland was 18.9%. Unemployment in rural areas was as high as 40%.
- Lower wages at around £3 per hour average.
- Lower gross domestic product (GDP) per capita. GDP of Poland is US$13,810 and GDP of UK is US$39,720.

Pull factors:

- Scotland was experiencing a skills shortage and needed more workers to fill vacancies in factories, farms, the service industry and in construction.
- Higher wages at, on average, £6 per hour, but depending on the type of jobs some workers were earning five times more than they would back in Poland.
- Plenty of low-cost flights are in operation between Poland and the UK.

Positive impact on Poland (donor country):

- Unemployment rate has fallen to 9%.
- Approximately £1.2 billion is sent from the UK to relatives in Poland every year (remittances), which helps them and the local economy.
- When Polish immigrants decide to return home, they can speak English; this makes them more desirable to employers in Poland.

Negative impact on Poland (donor country):

- The majority of emigrants are of working age, leaving some settlements where the majority of people left are children and old people.
- More males emigrate than females, resulting in a gender imbalance.
- Lots of highly educated and well-qualified people left Poland – this is referred to as a **brain drain** and leaves fewer working-age and well-qualified people in the donor country.

Positive impact on Scotland (host country):

- Polish is the second most commonly used language in Scotland and in the UK.
- Enriches the cultural diversity of Scotland.

- Increases the working population and reduces labour shortages.
- Polish immigrants are more willing to take on lower paid jobs, which may have unsociable working hours (service sector) or difficult working conditions (processing and manufacturing jobs). Scottish companies who employ Polish workers will benefit from a lower wage bill, making the company more competitive and their products/services better value for consumers.

Negative impact on Scotland (host country):

- Overcrowding in schools due to sudden increase in the roll.
- Money and resources needed for extra English literacy classes at school and also in the community for the adult population.
- Some GP surgeries and hospitals initially found it difficult to understand the Polish patients due to the language barrier.
- Overcrowding in the affordable parts of the city. Strain on the local housing stock.
- Language barriers may lead to misunderstanding, mistrust and friction between the Polish people and the local communities.
- During an economic downturn, Polish immigrants may become the target of frustration caused by higher unemployment rates.
- Immigrants are often blamed for 'undercutting' wages as they are willing to work for lower wages and often for longer hours.

Causes and impacts of forced migration

Case study: Syria

Characteristics

Much of the Middle East, including Syria, has been politically unstable since the so-called Arab Spring in 2010, when a series of protests, riots and civil wars broke out in many Arab countries.

Syria has been in a state of civil war since March 2011. Over 200,000 people have died as a result of the fighting between the Syrian government and the rebel groups.

Syrian city of Homs destroyed by civil war

According to UNHCR (United Nations High Commissioner for Refugees) 5.7 million people have fled Syria since the start of the conflict. Syrians first fled in vast numbers to neighbouring Lebanon, Turkey and Jordan. In addition, 6.5 million people are internally displaced, in search of a safe place to live.

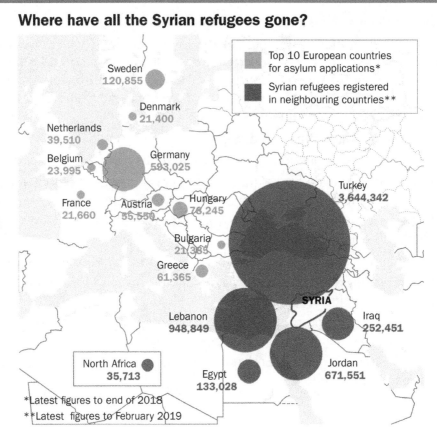

Where have all the Syrian refugees gone?

Top 10 European countries for asylum applications*

Syrian refugees registered in neighbouring countries**

Sweden 120,855

Denmark 21,400

Netherlands 39,510

Belgium 23,995

Germany 593,025

France 21,660

Austria 55,550

Hungary 76,245

Bulgaria 21,365

Greece 61,365

Turkey 3,644,342

SYRIA

Lebanon 948,849

Iraq 252,451

North Africa 35,713

Egypt 133,028

Jordan 671,551

*Latest figures to end of 2018
**Latest figures to February 2019

Estimated number and location of Syrian refugees, UNHCR

Incentives to move

Push factors:

- Thousands of Syrians flee their country every day. They often decide to finally escape after seeing their neighbourhoods bombed or family members killed.
- Bombings are destroying crowded cities.
- Horrific human rights violations are widespread.
- Thousands of people have been killed, kidnapped, tortured and raped.
- Basic necessities like food and medical care are sparse.
- Over 4 million people have been left homeless.
- Much of the country's infrastructure and essential services have been destroyed.
- Thousands of people are unemployed.
- For many Syrians it is impossible to lead a normal life.

Pull factors:

- Many Syrians have family and friends already living abroad so it is easier to make the move.
- Syria has porous land borders with Turkey, Lebanon and Jordan, which makes it relatively easy to migrate into these countries, even without a passport.

Impact on donor country (Syria):

- Much of Syria's educated elite population (those who have money and connections) have fled their home in search of safety.
- Syria does not have enough doctors and nurses to look after the injured and the weak.
- Once vibrant cities such as Homs and Aleppo are now ghost towns.

Impact on host countries (Lebanon, Turkey and Jordan) – now home to over 5.6 million Syrian refugees:

- The majority of Syrian refugees are living in Jordan and Lebanon. In the region's two smallest countries, weak infrastructure and limited resources are nearing breaking point under the strain.
- In some towns the population has doubled, putting a lot of pressure on health and education services.
- Waste management is not coping. Space is also an issue in crowded urban centres, where rents in some places have trippled since the influx of refugees.
- There are not enough teachers. Some schools send Lebanese children home at lunchtime and then teach Syrian children for the second half of the day.
- Some Lebanese people say they have lost their jobs because Syrians are willing to work for less, or that they have been evicted because Syrians share housing with many people, and can therefore afford rents that the Lebanese cannot.
- Since August 2014, more Syrians have escaped into northern Iraq at a newly opened border crossing. In a country that is still recovering from its own prolonged conflict, this influx is dramatic and brings additional challenges.
- An increasing number of Syrian refugees are fleeing across the border into Turkey, overwhelming urban host communities and creating new cultural tensions and resentments.

EXAM QUESTION

2012 saw a significant increase in Germany's population. This was not due to a sudden baby boom, but to the many immigrants moving to the country. Experts point out this could result in both benefits and problems.

Referring to a named case study, **analyse** the impact of migration on either the donor country or the host country.

6–8 marks

Rural

Rural land degradation in North Africa

North Africa – the Sahel region

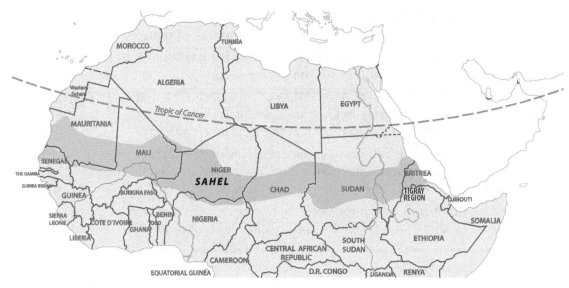

Map of the Sahel

The Sahel is a narrow band of land that stretches across North Africa. It is a semi-arid area that borders the southern edge of the Sahara. Many of the countries, e.g. Sudan, Eritrea, northern Ethiopia, in this region are susceptible to the process of desertification. This is the degradation of soil and vegetation cover and can occur in any of the dry areas of the world, not just desert fringes. 30% of the world's surface is under threat of desertification.

Desertification is defined as the spread of desert-like conditions in an arid or semi-arid area due to climate change or human influence.

Physical causes of desertification

Unreliable rainfall and drought: in the Sahel, rainfall is confined to just a few months in the northern hemisphere's 'summer', due to the arrival of the ITCZ. However, the total annual mean rainfall and the length of the wet season are both very unreliable. Periods of drought since the late 1960s are now more common as can be seen from the graph on the following page and this may be due to the ITCZ not migrating as far north in these years. It is thought this could be due to climate change.

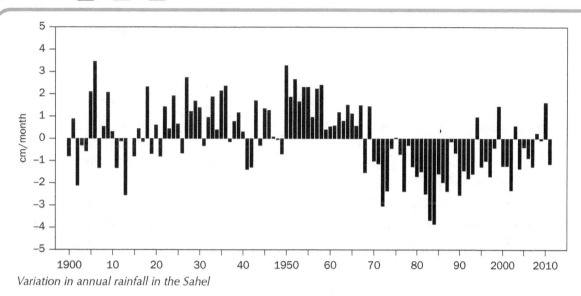

Variation in annual rainfall in the Sahel

Lowering of the water table: through prolonged drought and water abstraction from wells. This means certain plant roots can no longer reach a water supply.

Flash flooding: the rain often falls in a few heavy downpours and most is lost as surface runoff with very little infiltrating into the ground.

Water erosion: topsoil is washed away by the impact of raindrops (rainsplash), sheets of water (sheet wash), small surface streams that erode channels and deep gulleys (rill and gully erosion) that carve up the landscape.

Wind erosion: strong winds lift soil particles and blow them for up to several kilometres, creating dust storms. They are moved in three different ways:

- **Suspension**: very fine material (<0.15 mm) is picked up and carried large distances by the wind.

- **Saltation**: fine and coarse grain sand particles are lifted first vertically and then they are bounced in a leap-frogging action over each other.

- **Surface creep**: small stones and pebbles are dislodged by the moving sand grains and are rolled along the land surface.

Human causes of desertification

Population pressure: with an increasing population, the carrying capacity of an area can be exceeded, i.e. the maximum number of people that can be supported by the resources of the environment in which they live.

Deforestation: trees are cut down for firewood and building materials.

Overgrazing: the grazing of too many goats and sheep means the carrying capacity of the available grassland is exceeded and this leads to surface vegetation being stripped with little hope of it being re-established.

Children living in the Sahel region

Overcultivation: growing crops on the same areas of land for years without giving it a rest. This leads to a breakdown in soil structure, fertility and makes the land more susceptible to soil erosion.

Monoculture farming: growing the same crop for many years, which depletes the soil of moisture and nutrients.

Poor irrigation practices: farmers applying large amounts of water to their fields causes salts to leach out of the soil as it encourages the process of capillary action – salts in solution are transported through the soil horizons to the soil surface. Eventually this leads to soils that are uncultivable.

Cultivation of marginal land: fragile land that should not be used for crop growing.

Increased water extraction: with little or no surface water stores, more wells have to be sunk to tap into groundwater sources and this contributes to the lowering water table.

Consequences of desertification

Social	Economic	Environmental
Decrease in human wellbeing, e.g. poverty, famine, malnutrition.	Loss of crop production/yield.	Decrease in vegetation cover – pasture dries up.
Traditional nomadic way of life in Sahel areas is under threat.	Decline in biological productivity, e.g. growth of trees.	Increased soil salinity due to poor irrigation methods.
Increased rural–urban migration from areas suffering from desertification, particularly to towns and cities.	Animals, e.g. goats, sheep, cattle die due to starvation.	Increased dust storm frequency.
Families turn to 'famine foods', e.g. leaves in times of food shortages; these are nutritionally poor.	Stops development.	Increased surface albedo – this is the reflectiveness of a surface; with loss of vegetation, albedo increases as more incoming solar radiation is reflected back into the atmosphere.
Wells dry up – through over abstraction of water.		Decrease in biomass (amount of organic matter from living vegetation, i.e. leaf fall).
Political instability as a result of different 'tribes' coming together in urban areas.		Decrease in soil organic matter.
		Loss of key plant species.
		Loss of key animal species.

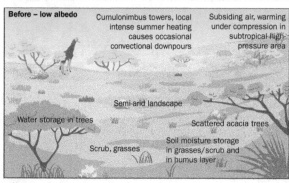 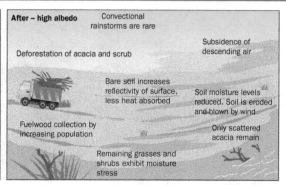

Effects of changing albedo on the landscape

TOP TIP

Just as in the River Basin Management topic, you should be able to divide the consequences of desertification into: social, economic and environmental.

Possible solutions to desertification

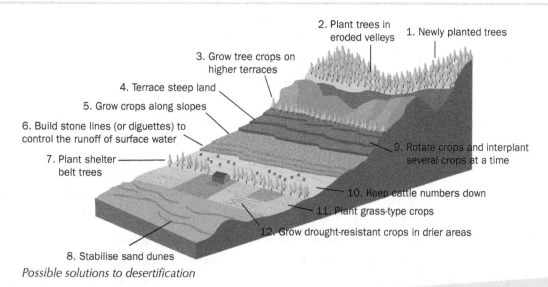

2. Plant trees in eroded velleys
1. Newly planted trees
3. Grow tree crops on higher terraces
4. Terrace steep land
5. Grow crops along slopes
6. Build stone lines (or diguettes) to control the runoff of surface water
7. Plant shelter belt trees
9. Rotate crops and interplant several crops at a time
10. Keep cattle numbers down
11. Plant grass-type crops
12. Grow drought-resistant crops in drier areas
8. Stabilise sand dunes

Possible solutions to desertification

Researchers have concluded that some of the most severe cases of land degradation in semi-arid areas could be reversed with the right policies and actions.

TOP TIP

Make sure you know all the physical and human causes of desertification and a number of solutions, and can explain them.

Case study: Tigray, Ethiopia

In regions like Tigray in northern Ethiopia, Africa, which is part of the western Sahel region, rural land degradation has been a major problem. It is an area of highland with a population that has increased tenfold since the end of the civil war and a population density of 100 people per sq km. The majority of the population are subsistence farmers and over the years the land has become badly degraded. Land degradation here is the result of population pressure, frequent droughts and poor farming practices.

Watch the video 'Regreening Ethiopia's Highlands: A New Hope for Africa' at **http://www.youtube. com/watch?v=nak-UUZnvPI**

Torrential summer rains cause soil erosion as topsoil is washed off the steep slopes. However, local people are gradually beginning to coax it back to recovery with help from experts.

Some of the management practices introduced include:

- The government terraced steep slopes and built stone bunds. These are stone walls that follow the contours of the hills and they prevent soil erosion and flooding.
- Grazing animals, e.g. goats are kept in enclosed areas near settlements and fodder is brought to them rather than letting them graze on open ground.
- Areas of extreme degradation were closed off to grazing, crop cultivation and tree felling.
- Forests were replanted.

Since these management practices were introduced in the 1980s, there has been a remarkable recovery of vegetation and also improved soil protection.

Although the government initiated the recovery, local communities came to recognise the value of such conservation work because they could see the benefits, such as reduced flooding and less soil loss.

Case study: The Great Green Wall

This is a long-term project to plant an 8,000 km belt of vegetation, trees and bushes across the Sahel region of Africa. It will eventually stretch from Senegal in the west to Djibouti in the east.

The aim of this wall is to combat desertification, land degradation and drought. The planted vegetation will green the land and protect the agriculture taking place around it. It is hoped that it will lead to the restoration of 100 million hectares of land, provide food security for 20 million people, create 350,000 jobs and offset 250 million tons of carbon by 2030. Progress is well underway in Senegal where 4 million hectares has already been restored. Large numbers of acacia trees have been planted by local communities. These trees will act as a shelter belt and reduce wind erosion; their roots will bind the soil together and reduce surface runoff and they will also provide shade, meaning the ground loses less water to evaporation. A secondary benefit has been an increase in wildlife returning to the region, e.g. hyenas, porcupines.

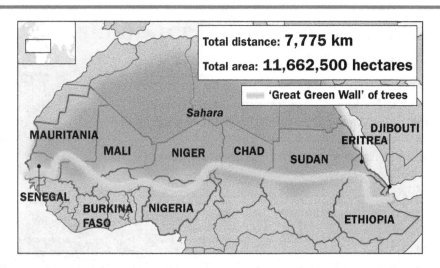

Total distance: **7,775 km**

Total area: **11,662,500 hectares**

'Great Green Wall' of trees

The estimated cost is US$8 billion and is being funded jointly by the World Bank, UN, African Union and UK Botanical Gardens.

For this project to be effective and completed, there needs to be good international coordination and local communities must feel part of the project and be able to see and believe in the benefits it will bring to their lives.

Watch the video 'Why is Africa building a Great Green Wall?' at **https://www.youtube.com/ watch?v=4xls7K_xFBQ**

EXAM QUESTIONS

1. a) **Explain** the purpose of building stone lines or bunds (or diguettes) as a strategy to stop land degradation in semi-arid areas.

   b) **Discuss** the effectiveness of another strategy, referring to a case study of your choice.

   12 marks

2. **Discuss** the social consequences of rural land degradation in a semi-arid area.

   10 marks

Rural land use conflicts in a glaciated landscape: The Lake District

The Lake District is England's largest national park and is home to Scafell Pike, Wastwater and thriving communities like Keswick and Bowness-on-Windermere. The area is enjoyed by over 15 million visitors every year due to its accessible location and beautiful scenery.

An aerial photograph of the Lake District

There are several conflicts that occur as a result of the large volume of tourists visiting the region.

Traffic congestion and parking issues

Conflicts	Solutions	Effectiveness
• The many visitors arriving at the Lake District often travel by car. There are bottlenecks in Bowness and Ambleside that often cause **traffic congestion**. • **Congestion** on the narrow, winding roads causes **access problems** for locals and the emergency services.	• **Boat shuttles**, e.g. Bowness to Hawkeshead on Windermere have been introduced to reduce congestion on the roads. • Proposals have been forwarded to reopen the Keswick to Penrith **railway** and to increase services on the Windermere branch line.	• This has been a success as the **Cross Lakes shuttle service** is growing in popularity and has been extended to other lakes in an effort to reduce traffic. • However, it would be **impractical financially** and also extremely difficult to extend the railway in the Lakes given the **relief of the land**.

Conflicts	Solutions	Effectiveness
• A **lack of parking facilities** means that cars often park on the roadside. This creates further congestion as traffic cannot freely move through the roads. Grass verges have also become severely eroded as cars are left at the roadside. • **Locals** conducting their everyday business often experience **problems such as getting from place to place** and crossing the road. • Pollution has reached dangerous levels in certain areas.	• **Grasmere now offers no on-street parking** to reduce congestion, and Keswick Main Street has been **pedestrianised** with car parks found on the edge. • Some villages have set up **restricted parking zones**, for example in Elterwater. The car park on the edge of the village has been expanded and parking on grass verges and near houses has been restricted. • **Public transport** has been **improved and subsidised**, for example the 'Langdale Rambler' bus service. Visitors are encouraged to use the buses instead of bringing their cars into the national park.	• **Traffic continues to be a problem**, especially in areas such as Grasmere, which has one of the highest levels of parking fines issued in the UK. This is due to the increasing number of visitors to the area but also the ever-increasing car ownership figures. This is an extremely difficult problem to overcome as there is only one train station in the area, which is far from the main village. They have had limited success as people prefer the convenience of their own vehicles.

Footpath erosion

Conflicts	Solutions	Effectiveness
• Many visitors use the footpaths in the area. They sometimes stray from the paths or walk on the edge, which leads to **footpath erosion**. In turn, this **scars** the landscape, e.g. Brown Tongue to Scaffel. • There are currently around **200 paths in need of repair** and this is an **expensive and time-consuming** job.	• **Tourist guides have been produced that do not include various areas** in order to reduce pressure on them, e.g. Brown Tongue to Scafell and Tarn Hoes. • 'Fix the Fells' is an initiative that repairs and maintains mountain paths. • Also, the Lake District National Park is working with landowners such as the National Trust to **provide pitched surfaces**, reseeding and construction of drainage channels.	• Due to the vast number of visitors, footpath erosion is an **ongoing problem** with approximately 200 paths in the Lake District in need of repair.  *Example of footpath erosion* • 'Fix the Fells' is a very successful initiative, but it relies on the generosity of volunteers.

Holiday homes and second homes

Conflicts	Solutions	Effectiveness
• Many houses and other types of accommodation have become **second homes** (around 20% in the Lake District), bought by wealthy people from the cities, and these can lie empty for a large part of the season. • This causes **house prices to inflate** and locals are then forced out as they cannot afford the costs. The Lake District National Park Authority has called the house price issue an 'acute' problem. • **Various local services such as primary schools** (Ulpha Primary) then **close** as those who own the second homes do not need to use the services.	• The **'Rothswaite Scheme'** (Borrowdale) attempted to ensure some properties were **available for locals** and not lost to second home owners. To qualify you had to have lived or worked in the area for at least three years.	• While the scheme was successful, it was **on too small a scale**.  *A holiday home in the Lake District* • Does not tackle low wage levels leading to a gap between local incomes and house prices.

Use of the lakes

Conflicts	Solutions	Effectiveness
• Watersports create waves, which in turn cause **erosion** of the shoreline. • Loud noises from the watersports can also **scare fish**, which means fishermen are often in conflict with those who participate in noisy water sports. • Some boats can cause **water pollution** and this affects those wanting to swim in the area.	• A **speed limit** was introduced on Lake Windermere.  *The lakes are used for many different water sports*	• Conservationists welcomed the new speed limit, but speedboat owners, water skiers and boat companies around the lake objected to the change. **Businesses have been affected** and boat users have had to find alternative lakes.

Urban

In this section

- the need for management of recent urban change (housing and transport) in a developed and in a developing world city
- the management strategies employed
- the impact of the management strategies

The urban population

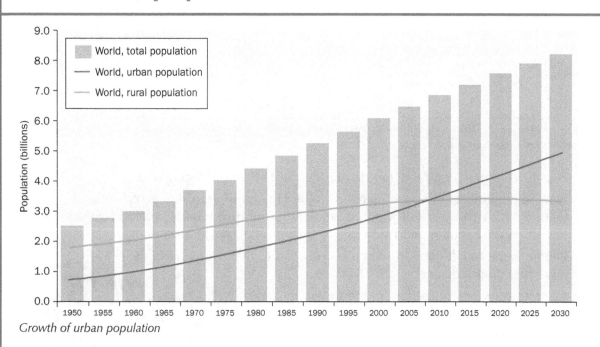

Growth of urban population

Urbanisation is the growth of towns and cities. In 2008, for the first time in human history, more people lived in the cities than in the countryside.

The status of 'megacity' is given to a settlement that has a population of 10 million and above. In 1975 there were three megacities (Tokyo, Mexico City and New York), now there are over 30 and most of them are in developing counties. General population and urban population are both growing most rapidly in developing countries and less so in developed countries.

All cities suffer from similar problems: pollution, traffic congestion, house price fluctuations, crime, demand for key services, demand for jobs, derelict land and urban decline.

Urban change and management in a developed world city: Glasgow

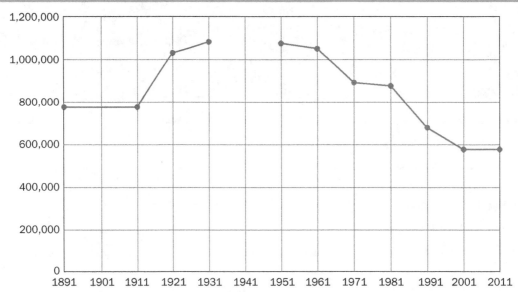

Population of Glasgow: census data

- Glasgow's population grew rapidly during the 19th and 20th centuries.
- New immigrants to Glasgow came from the Highlands and Islands of Scotland (early 1800s). This was followed by immigration of Irish (1840s), Italian (1890s), Jewish (WWII), Indian and Pakistani (1940s) and Polish (2000s) people.
- People came to Glasgow in search of jobs and to make a living in the city's booming industries.
- Most new immigrants settled in the inner city where rent was most affordable.
- The population of Glasgow peaked in the 1950s at 1.089 million.

TOP TIP

In an exam, name your case study city/area clearly at the beginning of your answer. Make sure your answer is detailed and not generic.

Changes to housing

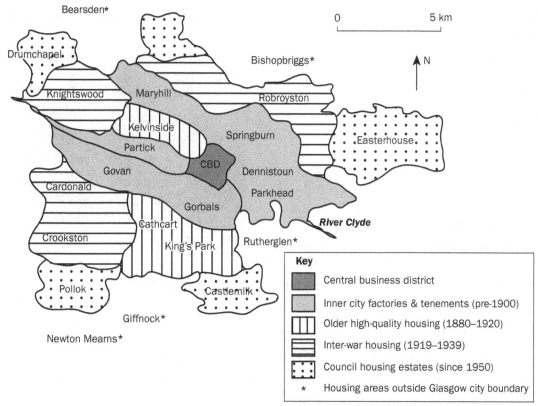

Inner-city areas of Glasgow

Housing change in the Gorbals area of Glasgow was necessary because:

- By the 1930s the Gorbals had a reputation as the 'slum of Europe'.
- There was serious overcrowding.
- Lack of sanitation – no indoor bathrooms in tenements.
- Poor state of repair.
- Poor living conditions lead to poor health – Glasgow had the highest mortality rate from tuberculosis (TB) in Europe.

Changes were made in the 1960s by knocking down old tenements and building new **high-rise flats**. These initially provided good accommodation but due to poor maintenance they were also demolished by the 1990s.

Construction of the high-rise flats

Most recent changes to housing: 'New Gorbals'

- New Gorbals has a mixture of private and social housing.
- Some old tenements were kept. Instead of demolition they were renovated to modern standards.
- New houses are mid-rise with 3–4 floors, containing 6–8 units of flats with each flat having two or three bedrooms.
- Larger 'town house' properties have been built to encourage families to stay in New Gorbals.
- The area is well served by NHS services, schools and public transport. It also has a library, sports centre and shops.
- Old industrial sites have been decontaminated and cleared. The land was then used to build houses, offices and parks.

New flats on Thistle Terrace, New Gorbals

For more on housing, see also *Leckie Higher Geography Course Notes*.

Changes to transport in Glasgow

There are many causes of traffic congestion, including:

Old and narrow streets

- Merchant City is the oldest part of the city and was built before the invention of cars so the streets are narrow. This problem is made worse by cars parking on the sides of the streets.
- There are many old historical buildings in Glasgow that cannot be knocked down to make wider streets.
- Streets close to George Square were laid out quickly during the Georgian period in a grid-iron pattern. This means traffic cannot flow smoothly due to there being many intersections in a relatively short space of time.

More cars on the road

- 69% of households in Scotland had at least one car or van in 2011 compared with 66% in 2001. In Glasgow at least 160,000 residents own a car.

Daily commute

- 49% of Glaswegians commute to work by car. The majority of people who work in the city centre do not live in the city – this creates morning and evening rush hours.
- Cities like Glasgow, with a river as part of the landscape, require many crossing points. Currently, Glasgow does not have enough.

Effects of traffic congestion

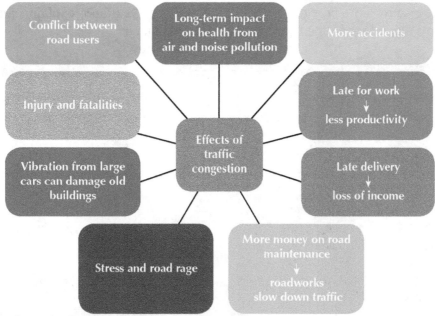

Effects of traffic congestion

Solutions to traffic congestion

Solutions	Effective?
One-way streets such as Hope Street and St Vincent Street.	Yes – traffic can move more smoothly **but** one-way systems can be confusing for drivers and it can take longer to reach your destination.
Motorway extension, e.g. M74 extension.	Yes – **but** it encourages more cars on the road, which leads to more pollution.
More traffic wardens, parking meters, build more multi-storey car parks.	Yes – it discourages people from parking illegally, which improves traffic flow. Parking fees can be expensive.
Build more bridges and tunnels and more footbridges (such as the 'squinty' and 'squiggly' bridges).	Yes – more bridge points between Southside of the city and the CBD will reduce congestion. More people now walk and cycle.
Improve public transport, Park and Ride systems, pedestrianise some streets.	Bus and cycle lanes make journeys on public transport quicker and people feel safer walking in CBD.

Solutions and effectiveness

EXAM QUESTION

For Glasgow, or any named developed world city you have studied, **explain** schemes that have been introduced to reduce problems of traffic management.

12 marks

Urban change and management in a developing world city: Kibera, Nairobi, Kenya

Nairobi is the capital city of Kenya, and is a city with a population of approximately 3.4 million people. The population of Nairobi has grown rapidly since the 1950s. 25% of Kenyans and 43% of the country's urban workers live in Nairobi.

Up to 60% of the total population of Nairobi lives in informal settlements known as shanty towns/slums. Kibera is one of the largest slums in Africa.

Need for management in Kibera

- Population of approximately 1 million.
- Massively overcrowded – the population density is 49,228/km².
- Average income is less than £1 a day, meaning over 50% of the population is living below the poverty line.
- Most people don't have a regular job. Two-thirds of the population work in the informal sector (black market).
- Up to 50% unemployment rate.
- Land is owned by the Kenyan government.

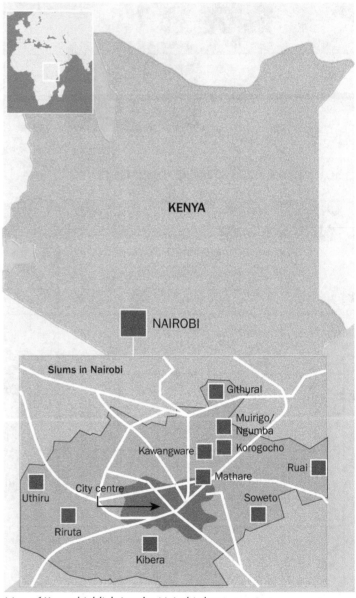

Map of Kenya highlighting the Nairobi slums

- Rent for a small tin shack is approximately £6 per month. 90% of the population are tenants with very little tenant security.
- 22% of the slum households have a water connection. There are two main water pipelines in the area and residents pay to collect water.
- No flushing toilets. One latrine is shared by 50–60 households.
- 20% of the population have electricity.
- No government-run clinics or hospitals.
- High crime rate.
- High number of street children and orphans.
- Many social problems related to drugs and alcohol.
- Very little help from the Kenyan government and Nairobi city council. Local officials and police force are often accused of being either uninterested in the slum problems or corrupt.
- There are no organised refuse collections. The land and water is heavily polluted.
- There is a lot of air pollution from the use of kerosene lamps and stoves due to lack of affordable electricity supply.
- High infant and child mortality rates, low life expectancy rate and recurring health problems: diarrhoea, malaria, TB.
- 60% of adult population are HIV positive.
- Low school enrolment and secondary school completion, especially for girls.

Management strategies employed in Kibera

- Slum upgrades by KENSUP (Kenya Slum Upgrading Programme).
- Micro-financing – a community-based financial service to help entrepreneurs start up businesses. Due to lack of official financial history, normal banks are unwilling to lend to people from the slums. Micro-financing companies like KIVA will lend the money to help people help themselves out of poverty.
- Rain water collection.
- Peepoo bag – a personal, single-use, self-sanitising, fully biodegradable toilet that prevents faeces from contaminating the immediate area as well as the surrounding ecosystem. After use, Peepoo turns into valuable fertiliser that can improve livelihoods and increase food security.
- 'Adopt a Light' campaign by public–private partnership.
- Privatisation of some services like water distribution and refuse collection.

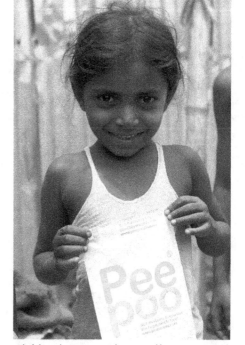

Child with 'Peepoo' bag – self-sanitising toilet

- Better arrangement of routes taken by minibus – same solutions for traffic congestion.
- Raising awareness and special campaigns.

- Educating the public about the Citizens' Rights Charter – to empower the slum residents and give them the tools to make effective demands on the city council.
- City council's budgets and accounts are now published to increase transparency.
- Involvement of charities, non-governmental organisations (NGOs) and community-based organisations to help improve aspects of life in Kibera, e.g. Water Aid for sanitation programmes.
- The Scoring Goals project. This project gives children and young adults the opportunity to play in organised football games and participate in drama competitions, building confidence and self-esteem.

Limitations of the management strategies

- Nairobi city council is understaffed and under-resourced to deal with the slum problems.
- Poor resource management and accounting.
- Weak and inefficient revenue collection (tax) by the government and city council.
- Lack of communication and dialogue between the residents and the city council.

Here is a link to an informative documentary about Kibera: **http://youtu.be/jQeKEGrDoQ4**

EXAM QUESTION

With reference to a named city in a developing country, **discuss** the social, economic and environmental problems often found in such cities.

12 marks

River basin management

In this section

- physical characteristics of a selected river basin
- need for water management
- selection and development of sites
- consequences of water control projects

TOP TIP

The concepts that you have learnt about in the Hydrosphere chapter will help you understand this topic, so make sure you revise it first and identify the links.

Global water surplus and deficiency

Management of the world's water resources is necessary as the hydrological cycle does not spread freshwater evenly over the globe. Some places have a surplus of water and other places have a water scarcity.

71% of the Earth's surface is water with 97% of that being salty and only 3% being fresh. Of this 3%, 2.25% is stored in ice caps and glaciers. This leaves only 0.75% of the Earth's water to be shared out among the world's growing population.

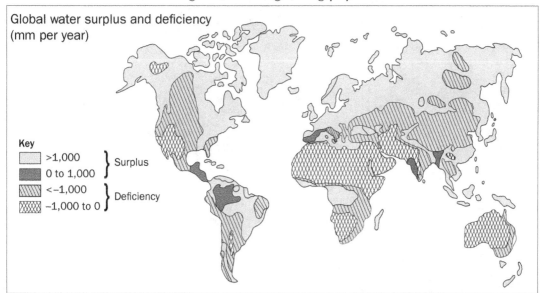

Global water surplus and deficiency (mm per year)

Key
- >1,000 } Surplus
- 0 to 1,000 }
- <−1,000 } Deficiency
- −1,000 to 0 }

Map showing global water surplus and deficiency

In areas with a water surplus, precipitation exceeds evapotranspiration and in areas of deficit, evapotranspiration exceeds precipitation.

River basin management (RBM) is all about controlling water surplus for human benefit and finding ways of creating a more reliable water supply for areas of deficit.

In the areas with a global water surplus, you will find the world's largest river basins, e.g. the Amazon River. In areas of deficit, higher evaporation rates may mean low precipitation is ineffective and surface rivers will only flow seasonally or not at all.

Study the maps below as you will need to be able to interpret this range of graphical information in order to be able to describe and explain the location of river basins in different continents. In this case it is North America.

Watch this video for an explanation of global water surplus and deficit: **http://www.youtube. com/watch?v=16CeJQU1XnA**

1,000 km

Major river basins of North America

Physical map of North America

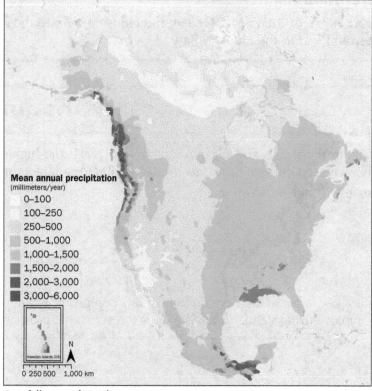

Rainfall map of North America

To describe and explain the location of the river basins in North America or any other continent, try to address the following points. The information below is again for North America.

General patterns of distribution and direction of flow

In North America, the Great Divide is a prominent watershed running from the Bering Sea southwards along the high peaks of the Rocky Mountains into Central America. Rivers on the east of this divide flow into the Atlantic or south into the Gulf of Mexico, e.g. St Lawrence and Mississippi rivers, and those to the west, like the Colorado River, flow into the Pacific. A spur of high ground comes off the northern end of this divide and runs east–west just north of the Great Lakes and all the rivers north of this flow into the Hudson Bay and Beaufort Sea, e.g. the Nelson and the Mackenzie rivers.

The distribution of drainage basins

An explanation of the distribution of drainage basins and drainage density or the number of rivers should refer to the presence of mountain ranges as major source areas of rivers due to having greater rainfall and possibly snowmelt in spring. Drainage density in North America is greatest in the east due to the annual precipitation being over 500 mm a year. Also, many of these rivers are fed by snowmelt in the spring, e.g. Missouri/Mississippi in their upper basins. To the west of the 500 mm isohyet, annual precipitation decreases and in the south-west of the USA, totals are as little as 25 mm a year.

Areas of high precipitation/water surplus and areas of low precipitation and high evaporation rates

Areas east of the Great Divide have high precipitation and lower evaporation rates resulting in a water surplus. Those west of the Great Divide have the opposite.

Position of river mouths

That is, the Colorado River to the Pacific Ocean; the Hudson River to the Atlantic Ocean; and the Saskatchewan River to the Hudson Bay.

Why the world needs water management

Human demands on freshwater as a result of population increase and higher standards of living have greatly increased through time. Freshwater is vital for a large range of activities:

- Domestic use, e.g. drinking, washing
- Agriculture
- Industry
- Recreation
- Hydroelectric power

Water availability is a key factor in the development of any country. Without a reliable supply it is very difficult to improve living standards and improve the economic prospects of a country.

How can river basins be managed?

River management involves interfering with the hydrological cycle. Many rivers across the globe do not have consistent flow throughout the year and so without control and regulation cannot be a positive and reliable resource. Historical records of discharge (such as a hydrograph showing discharge over many years) can show this uneven flow.

There are a number of different ways a river system can be managed for positive social, economic and environmental benefit:

- **Building a dam and creating a reservoir**: for water storage and for controlling a river's regime to reduce the risk of flooding, e.g. the Colorado River in south-west USA and Aswan High Dam in Egypt.
- **Irrigation channels for agriculture**: since the construction of the Aswan High Dam in the Nile Valley of Egypt, this fertile area can be farmed all year round, as it no longer relies on an annual flood to water the crops.
- **Transfer schemes**: moving water along aqueducts, through channels or pipes from an area of surplus to an area of deficit, e.g. from the Lake District to Manchester.
- **River diversion**: water is diverted to areas of deficit, e.g. California.

The hydrograph below shows the impact of the Glen Canyon dam (built in 1963) on the discharge of the Colorado River. Once a dam has been built, the hydrograph shows how the peaks and troughs in river flow have been reduced so the annual flow pattern is much more even. The advantages of this were:

- The dam reduced the size and frequency of floods.
- It reduced the amount of sediment that collected in the river, making the river, therefore, more efficient.

However, so that the previous river landscape and habitat are not lost altogether, periodic controlled flooding is sometimes used to kick-start the natural cycle of scour (erosion) and deposition of the river channel to create new sand bars or beaches and feeding areas for fish.

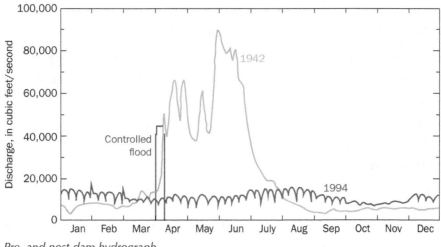

Pre- and post-dam hydrograph

Factors when considering the site of a dam

Physical	Human
Impermeable rock – so stored water does not seep away.	Low population so relocation is not too disruptive or costly.
Solid stable foundations on which the dam wall can be constructed safely.	Areas of no cultural or historical significance.
Large drainage basin with high drainage density and reliable supply of rainfall and/or snow melt.	Areas of poorer farmland where compensation to farmers is kept to a minimum.
Valley, gorge or narrow canyon – deep enough and with a small surface area (less water loss through evaporation) to store large quantities of water.	Close to agricultural areas so irrigation can be improved.
In areas not at risk of tectonic activity, i.e. unaffected by earth tremors.	Demand for electricity from hydroelectric power.

EXAM QUESTION

Discuss the physical and human factors that should be considered when selecting the site for a dam.

10 marks

Multi-purpose river basin management

These are large-scale schemes that address a number of differing demands:
- Water supply
- Regulation of river flow
- Flood control
- Hydroelectric power
- Navigation
- Recreation

Case study: The Colorado basin, USA

A multi-purpose scheme to provide a public water supply, flood control, expansion of irrigated farmland, power generation and recreation and tourism.

- Basin area: 637,137 km²
- Length of Colorado River: 2,334 km
- Average annual discharge: 637 m³/s
- Source: Rocky Mountains, Colorado
- Mouth: Gulf of California
- Average precipitation over basin: less than 25 mm
- Climate: semi-arid/arid
- Average precipitation in the basin: Upper Basin 1,600 mm (in Colorado and Wyoming); Lower Basin 84 mm (in south-west Arizona). Most of the Upper Colorado Basin precipitation falls as snow and melts in the spring, increasing river flow levels.
- Rock type: sedimentary (sandstone/ limestone)

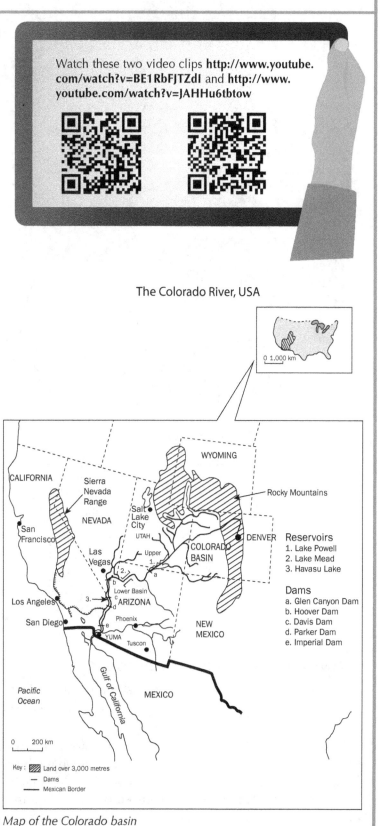

Watch these two video clips http://www.youtube.com/watch?v=BE1RbFJTZdI and http://www.youtube.com/watch?v=JAHHu6tbtow

The Colorado River, USA

0 1,000 km

Reservoirs
1. Lake Powell
2. Lake Mead
3. Havasu Lake

Dams
a. Glen Canyon Dam
b. Hoover Dam
c. Davis Dam
d. Parker Dam
e. Imperial Dam

Key: Land over 3,000 metres
— Dams
— Mexican Border

Map of the Colorado basin

Why was management needed?

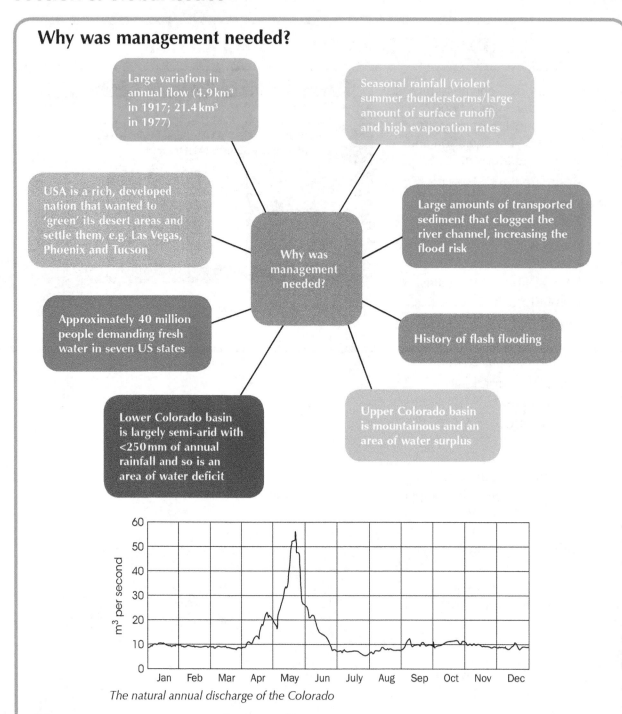

Large variation in annual flow (4.9 km³ in 1917; 21.4 km³ in 1977)

Seasonal rainfall (violent summer thunderstorms/large amount of surface runoff) and high evaporation rates

USA is a rich, developed nation that wanted to 'green' its desert areas and settle them, e.g. Las Vegas, Phoenix and Tucson

Why was management needed?

Large amounts of transported sediment that clogged the river channel, increasing the flood risk

Approximately 40 million people demanding fresh water in seven US states

History of flash flooding

Lower Colorado basin is largely semi-arid with <250 mm of annual rainfall and so is an area of water deficit

Upper Colorado basin is mountainous and an area of water surplus

The natural annual discharge of the Colorado

How was management imposed?

By building a series of large dams, starting with the Hoover Dam in 1935 and the creation of several reservoirs, e.g. Lake Mead behind the Hoover Dam and Lake Powell behind the Glen Canyon dam.

There were political complications: the Colorado flows through seven states so the Colorado River Compact was drawn up to give each state a water allocation. This was not finalised until the mid-1960s as it was so controversial; it is still contentious today with California and Arizona being accused of exceeding their water allocation.

Any of the RBM questions will want you to break down river management case studies into the social, economic and environmental advantages and disadvantages of the scheme. You should pre-prepare this breakdown for your individual case studies. The table shows the breakdown for the Colorado River.

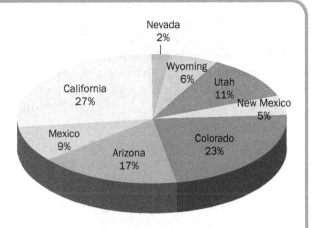

Colorado River allocation state by state

	Advantages	Disadvantages
Social	Increased employment from hydroelectric plants, e.g. Glen Canyon and industrial expansion. Increased recreation and tourism on Lake Powell, e.g. from houseboats, water skiing, kayaking. Urban water supply. Increased colonisation of the arid/semi-arid region of south-west USA. Phoenix is a boom city with some of the richest farmland in the USA due to the creation of irrigation canals.	Mexico in the Lower Colorado basin suffers by being at the end of the line with much of the Colorado having 'dried up' and prone to salinisation as, due to high evaporation rates, the river water is highly saline by the time it reaches the Mexican irrigation channels. Crops then suffer. Politics play a big part in river management in the US. Water projects can secure votes for Senators, and richer, more influential states will have more leverage when it comes to allocation. Native Indian burial grounds flooded.

	Advantages	Disadvantages
Economic	Provides a reliable water source for agricultural, industrial and public need in the south-west USA. Irrigation for farmers in the drier states of Utah, Nevada, Arizona and California. Imperial Valley in California is one of the largest irrigated areas in the world, growing water-hungry lettuces and alfalfa. Large cities such as Los Angeles and San Diego demand a huge amount of water. 20 dams like the Hoover, Glen Canyon, Davis and Parker produce a large amount of electricity. The Parker Dam uses this power to pump water along the Colorado Aqueduct to farms in southern California and San Diego city. Reservoirs, e.g. Lake Powell, are a tourist attraction, offering boating and other water sports, and this generates jobs and income.	Farmers pay a very low price for their water so are very wasteful. Industry using the cheap electricity is often very polluting, e.g. steel mills in Page, Arizona.
Environmental	Flood control. More even annual discharge so less erosion. Decreased sedimentation in river channel. Reservoirs provide a habitat for wildlife and birdlife.	Destruction of wilderness areas. Trapping of silt behind the dams so sand bars and islands do not form and valuable habitats downstream disappear. Large reservoirs lose millions of cubic meters of water through evaporation every year. Loss of desert wildlife when canyons were flooded to form reservoirs.

EXAM QUESTION

Explain why there is a need for water management in a river basin you have studied.

10 marks

Development and health

In this section

- validity of development indicators
- differences in levels of development between developing countries
- a water-related disease: causes, impact, management
- primary health care strategies

Indicators of development

TOP TIP

When referring to a developing country, don't use 'Africa' as a collective for all poor countries. Africa is a continent with 54 diverse countries.

Development is the use of resources to improve the standard of living of a country. Various social and economic **indicators** are used to measure the rate and level of progress. The world's richest 1% of the population owns 50% of the world's wealth. This shows that development is uneven in the world.

Useful terms

High-income countries	Low-income countries
Developed countries	Developing countries
EMDC: economically more developed countries	ELDC: economically less developed countries
e.g. UK	e.g. Tanzania

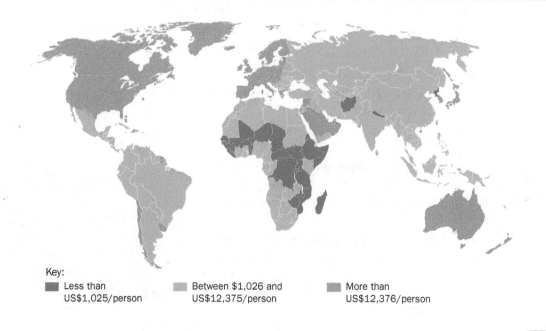

Key:

■ Less than US$1,025/person   ■ Between $1,026 and US$12,375/person   ■ More than US$12,376/person

The map on the previous page created using data from the World Bank: **https://datahelpdesk. worldbank.org/knowledgebase/articles/906519- world-bank-country-and-lending-groups**

Relationship between social and economic indicators

Economic indicators — Standard of living	Social indicators — Quality of life
Economic indicators show how well off a country is in terms of money, the currency usually used is US$. Per capita means per person.	Social indicators show how well a country is developing in key areas such as health, education and diet.
Examples of economic indicators: • Gross National Product (GNP) per capita • Gross Domestic Product (GDP) per capita • Gross National Income (GDI) per capita • Vehicles per thousand (‰) • Telephones ‰ • TV sets ‰ • % of people employed in agriculture • Energy consumption per capita (kw) • % of population living below the poverty line	Examples of social indicators: • Life expectancy • Birth rate • Infant mortality • Adult literacy • Number of patients per doctor • Number of calories per person per day • Gender equality • Secondary school enrolment

In most cases, if a country is creating a lot of wealth (high GDP) then it will in general also have a good quality of life (long life expectancy): wealthier countries usually have healthier populations.

However, some countries are an exception to this rule.

TOP TIP

Using the figures in the below table, you could plot a scatter graph to show the relationship between wealth (GDP per capita) and health (life expectancy).

Country	GDP per capita US$	Life expectancy	Adult literacy %	HDI* rank
Cuba	7602	79	99.8	73
India	1940	65	74.4	130
Jamaica	5110	76	87.9	97
Japan	38428	84	99	19
Tanzania	936	66	69.4	154
Turkey	10541	76	95.3	64
UAE	40699	77	77.9	34
United Kingdom	39720	81	99	14
Vietnam	2343	76	94	116
Source:	World Bank	WHO	UN	UN

*Human Development Index

Limitations of single indicators

The above indicators are examples of **single indicators**. They show one aspect of how the country is performing, and are very useful in giving a general idea of how the country is doing. However, there are limitations to these single indicators:

- Most indicators are an average figure. For example, GNP per capita is the average income of a person in a country if all of the wealth created in that country is divided equally among the population. GNP per capita does not show the inequality of wealth within a country. GNP figures are in some cases inflated by oil revenues.

- The standard of living and quality of life is generally better in the cities (because wealthy people tend to live in cities). Single indicators do not show regional variation (rural/urban), or account for other differences and inequalities, such as gender.

- Single indicators do not take into account the reasons for higher energy consumption in hot/cold countries.

- Economic indicators such as GDP/GNP/GNI per capita use US$ as a unit. This does not take into account the amount of money lost in converting a different currency into US$. It doesn't take into account the purchasing power of US$ in the local economy either, so it gives a false image of how rich/poor a country is.

- Subsistence agriculture and the information economy (bartering system) are not reflected in economic indicators.

- Some of the indicators will not apply to a country, e.g. if a country does not have widespread electricity then the number of TVs per household will be largely irrelevant.

- Data used to calculate the above single indicators will also have a degree of inaccuracy for the same reasons as inaccuracies in census data (see page 59 in the Population section).

One indicator is not enough to show how developed or underdeveloped a country is. In an attempt to have a more comprehensive indicator, the United Nations uses a combined indicator called the **Human Development Index**, which is calculated by combining the adult literacy rate, life expectancy and real GNP to give a more balanced view.

TOP TIP

Learn plenty of real country examples to include in your exam answers.

Inequalities between developing countries

Out of 224 countries in the world, over 100 of them are considered to be 'poor', i.e. developing countries. However, not all developing countries are equally poor.

Reasons behind global variations in development	
Physical factors:	**Human factors:**
• Climate	• Population growth
• Relief	• High levels of disease
• Resources	• Lack of industralisation and infrastructure
• Environment	• Trade/trade barriers
• Natural disasters	• Debt
• Geographical location	• Civil wars/government stability
	• Corruption
	• Education/technology

EXAM QUESTION

Referring to named developing countries that you have studied, **account for** the wide range in levels of development between developing countries.

10 marks

Malaria: water-related diseases

What is malaria?

- Malaria is an example of water-related disease found in tropical and sub-tropical areas of the world.
- About 3.3 billion people – half of the world's population – are at risk of malaria.
- In 2017, there were about 219 million malaria cases.
- Annually, there are approximately 445,000 malaria deaths.
- It is an infection caused by the malaria parasite called Plasmodium.
- It is a preventable and treatable disease.

The female Anopheles mosquito is the vector/carrier of the disease

The physical and human causes of malaria

Physical	Human
Female Anopheles mosquito (vector/carrier)	High population density (blood meal)
Stagnant water – puddles, paddy fields Correct environment for female mosquito to lay her eggs	Shade for mosquito to rest
Moderate to high rainfall	Poor sanitation in shanty towns can contribute to areas of stagnant water
Humidity over 60%	Dams, reservoirs and irrigation channels
Altitude below 3,000 m (warmer)	
Temperatures 15–40°C	

Symptoms

Symptoms usually appear 9–14 days after infection and include fever, shivering, vomiting and other flu-like symptoms.

If not treated, malaria can be deadly; early, accurate diagnosis (blood test) and treatment is vital.

Impact on the people and community

Malaria has serious economic impacts in Africa, slowing economic growth and development and perpetuating the vicious cycle of poverty.

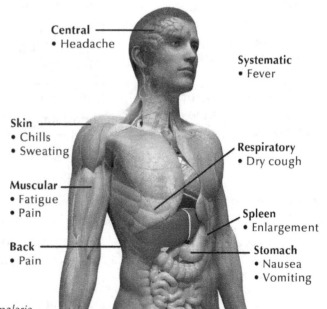

Central
- Headache

Systematic
- Fever

Skin
- Chills
- Sweating

Respiratory
- Dry cough

Muscular
- Fatigue
- Pain

Spleen
- Enlargement

Back
- Pain

Stomach
- Nausea
- Vomiting

Symptoms of malaria

Malaria is truly a disease of poverty, afflicting mainly the poor who tend to live in malaria-prone rural areas in poorly-constructed dwellings (shanty towns) that offer few, if any, barriers against mosquitos.

- Immune system can be weakened by malnutrition or other illness.
- If you are ill you are unable to work for at least 10 days, resulting in loss of income.
- Income is on average 60% lower than in non-malarial areas.
- The risk of infection is highest in the rainy season. This coincides with the agricultural peak, meaning that less food may be produced, as people cannot work.
- Chronic absenteeism in school children.
- Growth of GDP is lowered by 1–3% per year.
- Accounts for 30–50% of all hospital admissions.
- Costs up to 40% of public health expenditure.

Management

Prevention	Effectiveness
Spray insecticides – DDT	• Cheap and easy to apply but it is bad for the environment • Health risk • Banned in 2007
Spray insecticides – Malathion	• Less risky than DDT but it is expensive • It needs to be re-applied more often • Unpleasant smell
Throw mustard seeds into stagnant water – mosquito larvae get stuck to the sticky seeds and drown	• It works in small controlled areas but it is impossible to do this to all stagnant water • A waste of food
Introduce natural enemies – mosquito-eating fish like muddy loach and carp. Ducks can be released into the paddy field.	• It doesn't spoil the environment • People can eat the fish/duck – extra protein • It has been successful in parts of India and East Africa (Tanzania)
Drain breeding sites by filling depressions (holes) and planting eucalyptus trees	• Removes the breeding sites • Almost impossible to implement as the mosquito can breed in a muddy footprint • Canals need to be flushed every 5–7 days to disrupt breeding cycle – clean water is too valuable
Parasitic wasps – eat larvae	• No real danger to humans but may cause harm to the indigenous wildlife
BTI bacteria are grown on coconuts, which are then thrown into stagnant water and destroy the stomach lining of larvae	• Cheap, no risk to environment • Coconuts are plentiful and often grow near stagnant water • Lasts up to 45 days • Only works in large area – not a puddle

Treatment	Effectiveness
Chloroquine	• Cheap and easy to use • Some mosquitos have developed resistance
Lariam	• Lots of side effects
Malarone	• 98% effective • Few side effects • Expensive
Quinghaosu – developed from Chinese herbal medicine (commercial name is Artemisinin)	• Fast acting • Mosquitos not becoming resistant so far • Very expensive
RTS,S – new vaccine	• Triggers the human immune system to defend the body against the malaria parasite as it enters the bloodstream • Vaccine pilot launched in Malawi in April 2019 • GlaxoSmithKline (drug company) is trying to keep the cost of this drug down
Insecticide-treated bed nets	• Lasts up to one year • Reduced incidence in Tanzania by 35%
Insect repellents and covering the skin at dawn and dusk	• Insect repellent can smell unpleasant • Covering the skin is not always practical
WHO 'Roll Back Malaria' campaign, and the Bill and Melinda Gates Foundation	• Good, but billions of dollars must be raised

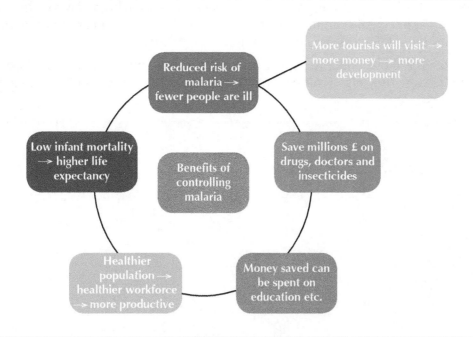

Improving health care in developing countries

The World Health Organisation (WHO) was set up by the United Nations (UN) in 1948. Since then the WHO has been helping poor countries to obtain better health care. The WHO is funded by the UN.

Their main aims are to eradicate diseases with the use of:

1. Vaccination

2. Quarantine

3. Research and development into cure and prevention of infectious diseases

4. Education and better access to health care

Example of WHO's work: primary health care (PHC)

Health care in many developing countries is so basic that any improvement (however small) will have a positive impact on the overall primary health care.

- Primary health care is the most basic level of health service.
- Its aim is to improve access to health care for EVERYONE in the country. This means providing basic health care to rural villages and settlements far into the bush lands.
- Primary health care is generally better in the city, but in the slums (such as Kibera, see page 88 in the Urban section) health care and living conditions are rudimentary.

Barefoot doctors (also known as 'community health workers')

A barefoot doctor is someone in the rural village/community who received minimal basic medical and paramedical training. Their purpose is to bring health care to rural areas where urban-trained doctors would not settle.

They promote basic hygiene, preventive health care, and family planning, and treat common illnesses. They also refer patients on to a doctor and hospital if the condition is serious.

They use cheap, cost-effective treatments such as oral rehydration therapy (ORT). ORT is a cheap and effective way to treat dehydration and diarrhoea with a solution of water, salt and sugar. Diarrhoea kills about 2.2 million children every year.

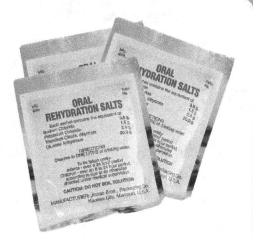

Other examples of PHC

- Advice on diet and nutrition.
- Advice on food preparation.
- Family planning: giving more choice to women about reproduction and contraception.
- Basic sanitation education, e.g. washing hands and boiling water.

Oral rehydration salts to treat dehydration and diarrhoea

- Colourful information posters and wall murals about how to prevent catching diseases; they usually contain very little writing because of the high level of illiteracy in developing countries.

Health education schemes

- Linked to local schools and volunteer mothers' groups – targeting suitable groups of people. Expensive but beneficial.
- Raise awareness of general health problems, the benefits of better diets and nutrition.

Mass vaccinations

- Offer protection against health problems that can kill in a developing country, such as polio, measles, whooping cough, cholera and TB.

Bamako initiative (Mali)

- 34 countries linked their hospitals, rural social services, schools, religious groups, women's and youth groups to develop PHC.
- In Benin, 200 health centres were set up covering 58% of the population. Costs were spread out among local communities.
- In 1993, a vaccination programme was introduced.
- Pre- and post-natal health education set up to tackle infant mortality rates.

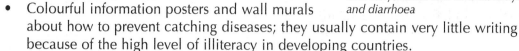

EXAM QUESTION

Referring to specific primary health care strategies you have studied, **evaluate** their effectiveness in meeting the needs of people in **developing** countries.

10 marks

Global climate change

In this section

- physical and human causes
- local and global effects
- management strategies and their limitations

'Climate is what you expect – weather is what you get'
Robert A. Heinlein

Physical causes of climate change

The Earth's climate has always varied. The main physical factors influencing temperature variations are explained below:

Milankovitch cycles: global temperatures change due to variations in the tilt and orbit of the Earth. This alters the distance the Earth is from the sun, which leads to changes in the amount of solar insolation received. The three dominant cycles are known as **eccentricity, axial tilt** and **precession.**

Volcanic eruptions: large volcanic eruptions (e.g. Eyjafjallajökull in Iceland) can release vast quantities of ash and dust into the atmosphere. This ash and dust can then absorb and scatter insolation, which temporarily lowers temperatures. Moreover, volcanoes release sulfur dioxide, which creates fine sulfate aerosols in the stratosphere. The reflection of solar radiation is further increased. Although volcanic eruptions increase the amount of carbon dioxide (a greenhouse gas) in the atmosphere, which can lead to longer-term warming, the amount of carbon dioxide released by volcanoes only accounts for less than a percent of that released by human activities.

Sunspots: an increase in solar activity (sunspots) may lead to an increase in temperature. Solar activity tends to occur in cycles of 11 years, but their influence is regarded as minimal in the scientific community.

Retreating/melting ice caps: melting ice caps can expose the land underneath, which reduces the albedo of the Earth's surface. This can result in an increase in temperatures and is known as a positive feedback loop.

Eccentricity   100,000 year cycles

Aphelion   Sun   Perihelion

Obliquity (axial tilt)   41,000 year cycles

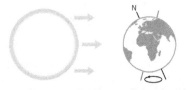

22.1° 0°
24.5° N

Precession   26,000 year cycles

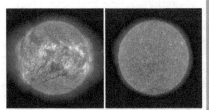

2002 solar activity on the left versus 2009 solar activity on the right

Melting ice caps can also decrease water temperatures and salinity in the surrounding seas and oceans. This can alter currents such as the North Atlantic Drift, which accounts for nearly one-third of Western Europe's heat.

Human causes of climate change

'The latest Assessment Report of the Intergovernmental Panel on Climate Change (IPCC) states that warming of the climate system is unequivocal, and that it is *extremely likely* (>95% confidence) that human influence has been the primary cause of warming.' (Met Office, UK)

There is a wealth of information on climate change at the BBC site here **http://www.bbc.co.uk/education/guides/z432pv4/revision/3**

Since the start of the Industrial Revolution in the late 18th century, the concentration of greenhouse gases in the atmosphere has increased significantly. This has been linked with a phenomenon known as the 'enhanced greenhouse effect' that increases global temperatures and causes the climate to change. Some of the human causes of climate change are explained below:

1. Increase in carbon dioxide emissions

- **Electricity generation**: Burning fossil fuels to generate electricity in power stations has led to a higher concentration of carbon dioxide in the atmosphere. There has been a rapid increase in the global demand for power due to an increased population, technological advancements and economic development, and therefore global emissions from energy production have increased significantly in recent decades.

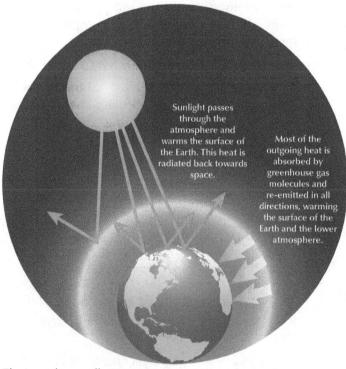

Sunlight passes through the atmosphere and warms the surface of the Earth. This heat is radiated back towards space.

Most of the outgoing heat is absorbed by greenhouse gas molecules and re-emitted in all directions, warming the surface of the Earth and the lower atmosphere.

The 'greenhouse effect'

2. Reduction in carbon sinks

- **Deforestation**: The removal of vegetation has taken place for activities such as agriculture, mining and urban development. Trees convert CO_2 to oxygen, but the removal of this 'carbon sink' is detrimental for the atmosphere.

- **Peat bog reclamation or development**: Peatlands are one of our most valuable ecosystems and are the largest natural terrestrial carbon stores on the planet. When peat comes into contact with the atmosphere, its carbon combines with oxygen and releases carbon dioxide. Peatlands often contain vegetation that acts as a natural carbon sink. The development of peatlands damages the land and reduces vegetation cover. Many of Scotland's windfarms are constructed on highland peat.

3. **Increase in methane emissions**
   - **The decomposition of waste in landfill sites**: Methane is released from decomposing waste – this accounts for almost one-third of the UK's methane emissions.
   - **Agriculture**: An increase in global population, economic development and associated demand for food has led to increased emissions from the agricultural sector. Rice paddies and cattle farms release high levels of methane.

4. **Increase in nitrous oxide emissions**
   - **Vehicles**: Globally, the number of vehicles is increasing due to a rising population and increased wealth. Nitrous oxides are dangerous greenhouse gases (GHGs) that are emitted from vehicle exhausts.

5. **Chlorofluorocarbon (CFC) emissions**
   - **Older refrigeration and air-conditioning units**: CFCs are emitted from older air-conditioning systems and refrigerators and can remain in the atmosphere for 400 years. The use of CFCs is highly regulated due to its potency and damaging impact on the ozone layer.

6. **Increase in hydro-fluorocarbon (HFC) emissions**
   - **Newer refrigeration and air-conditioning units**: HFCs are potent greenhouse gases used to replace CFCs in refrigeration and air-conditioning units. They have now been found to have a warming potential thousands of times more powerful than CO_2 and their use is to be phased out over the next decade.

EXAM QUESTION

A study found that 16% of species in the world would face the risk of extinction solely because of climatic factors. Based on current climate projections, global average temperatures will reach about 4°C higher than pre-industrial times by 2100.

Both human and physical factors contribute to climate change.

Explain the physical causes of climate change.

TOP TIP

Bring a highlighter into the exam. Highlight key words in the question. In this case you should only write about the **physical** causes of climate change.

8 marks

Local impact of climate change

TOP TIP

There are lots of figures and statistics in this section. Remember you do not need to learn them all. Pick a few that stand out!

Glasgow City Council consider climate change to be one of the most serious challenges facing Scotland today.

Significant trends have been identified in West Central Scotland.

1. Overall increase in average rainfall and more intense rainfall events

- **Increased incidences of flooding:** This causes water damage to infrastructure and buildings – the RAH Hospital in Paisley, the M74 and the M8 motorway are particularly vulnerable according to the 'Climate Ready Clyde' review.

- **Increased cost of flood management:** More expensive, local flood defence schemes are needed due to increased risk of flooding. Network Rail states that it spends £25m a year in Scotland to protect its lines from floodwater.

Flood defence scheme in Houston, Renfrewshire

- **Damage to traditional stonework:** Traditional buildings will be wetter for longer periods of time, resulting in increased weathering of stone. This puts many of Scotland's historic buildings at risk of decay.

2. Increased maximum temperatures and length of growing seasons

- The length of the growing season in the west of Scotland has increased by 5 weeks since 1961, which has increased crop yields. However, diseases such as potato blight may be more difficult to control, and some non-native pests and animal diseases may become more established.
- The longer growing season will have a detrimental impact on hay fever sufferers.
- The increased temperatures will prove detrimental for species such as the Arctic char (fish) which cannot adapt to the speed at which the temperature is changing.

3. Rising sea level of up to 2mm a year

- A rise in sea level will impact on the coastlines of Scotland, destroying salt marshes and low-lying coastal settlements. Glacial isostasy (the upward movement of the Earth's crust since the end of the last glacial maximum) should mean that the impact in Scotland is reduced.

4. More frequent and intense heatwaves

- An increase in summer heatwaves in Glasgow may eventually cause increased mortality, particularly affecting vulnerable groups. This will put pressure on the local health care systems.
- The hottest June day on record (31.9°C in June 2018) led to Glasgow's Science Centre roof melting.

5. Milder winters

- Milder temperatures in winter will mean even fewer snow days in the future. There has already been a reduction in snowfall over the last century, but it is predicted that winter snowfall may reduce by 50% or more across Scotland by the 2080s. This will negatively affect the ski industry in Scotland.

For more on how Scotland's climate is changing, take a look at this website: **https://www.nature. scot/climate-change/climate-change-impacts- scotland**

Global impact of climate change

Climate change has global ramifications. However, a report from The Millennium Project states: 'Poorer countries that contribute the least to GHGs are the most vulnerable to climate change's impacts because they depend on agriculture and fisheries, and they lack financial and technological resources to cope.'

Some of the global implications of climate change are explained below:

1. Sea level rise

An increase in global temperatures, particularly over the past century, has resulted in glacial retreat and a reduction in land ice cap cover. This, combined with the thermal expansion of the warming oceans, has led to an increase in global sea levels – they have risen by 20 cm over the last century.

- Low-lying coastal countries, such as Bangladesh, will be affected most severely by rising sea levels, with large-scale displacement of people, loss of land for farming and destruction of property.
- Sea level rise poses an existential threat to low-lying island nations such as the Maldives (which has an average altitude of only 1 m above sea level).
- Some governments and land-owners have had to invest greatly in expensive flood protection schemes. The city of Miami in Florida has spent $500 million to elevate infrastructure and purchase pumps to divert the rising sea water.

2. Increased intensity of extreme weather events

Hurricanes, tornadoes and storms have increased in intensity and are predicted to continue to get stronger.

- Hurricanes can cause deaths, infrastructure damage and large-scale displacement of people (e.g. Hurricane Katrina). In some coastal locations of the USA it is almost impossible for locals to insure their homes.

3. Exacerbation of dry conditions

- A prolonged dry period can lead to wildfires (e.g. in California) which can in turn lead to deaths and property damage.
- Potential conflicts over water supplies have and will continue to arise. One of the contributing factors leading to the war in Syria was drought, which led to unequal access to water and mass rural to urban migration.

4. Increase in vector-borne diseases

- More people will be at risk from vector-borne diseases such as malaria. The Anopheles mosquito will move into new areas made habitable by changing temperatures and precipitation levels.

5. Ecosystems are changing

- Sensitive coral and algae that live on coral are starved of oxygen, causing bleaching, e.g. the Great Barrier Reef off the coast of Australia.
- Potential extinction or reduction in the number of certain species. It is estimated that up to 60% of the current Adélie penguin habitat in Antarctica could be unfit to host colonies by the end of the century.

Adélie penguins on Antarctica

TOP TIP

Embed examples throughout your answer.

EXAM QUESTION

Discuss a range of possible climate impacts.

10 marks

Management strategies and their limitations

Some of the management strategies used to mitigate against climate change are discussed below.

Strategy	Explanation	Successes and limitations
Carbon capture and storage	Reduces greenhouse gas emissions from fossil fuel-based electricity production by preventing carbon gases from entering the atmosphere. e.g. at Peterhead, Aberdeenshire.	**Successes:** • Reduces carbon emissions, in some cases by 90%. **Limitations:** • Long-term security of underground storage is difficult to predict and there is a risk of the stores leaking.
The use/ increased use of renewable energy sources (e.g. wind and geothermal)	**Harnessing wind energy e.g. Whitelee Wind Farm** Wind turns propeller blades that spin a rotor. This in turn spins a generator to produce electricity.	**Successes:** • Wind is clean, free and renewable. • Employment opportunities – in the US alone, 100,000 people are employed in the wind energy sector. **Limitations:** • Some people oppose wind farms because they regard them as unsightly and noisy. • Prime sites for wind turbines are often in remote locations away from the demand. This results in the costly development of long transmission lines.
	Geothermal power supply e.g. Iceland This is where the natural heat of the Earth is used to produce steam and power turbines.	**Successes:** • Small land footprint. • Can generate a lot of power. **Limitations:** • Expensive to construct. • Potential issues with water pollution (heavy metals in the reservoirs).
Climate engineering (such as solar radiation management)	A type of geo-engineering project that aims to reduce the amount of sunlight hitting the Earth and thus controls temperatures. This may be achieved by whitening clouds or injecting particles into the stratosphere.	**Limitations:** • This does not address the issues of an increased concentration of greenhouse gases in the atmosphere and associated problems such as ocean acidification. • Long-term effects have not been studied.

A multi-faceted approach to reducing vehicle emissions	**Congestion zones and associated charges** In cities such as London, drivers are charged for entering the congestion charging zone. This is to discourage drivers from entering the central zone during peak hours and promotes the use of public transport, thus reducing emissions.	**Successes:** • Proven effective in reducing traffic congestion and air pollution in urban areas. **Limitations:** • Quite unpopular among the general public. • Cities need a robust public transport system to accommodate this. • Small scale and localised.
	Other charges and grants Some governments (e.g. the UK) provide grants towards electric cars and tax 'gas guzzlers'. They also promote small changes such as fuel-efficient driving. This reduces emissions per vehicle.	**Successes:** • Also tackles the issue of air pollution (and associated health implications). **Limitations:** • Electric cars remain expensive and do not yet offer the convenience of a diesel or petrol car. Charging stations are required. • The source of electricity to power the electric car needs to be considered – is it from a coal-fired power station? • 'Driving smarter' is small scale and more is needed to tackle emissions.
The Paris Agreement	The Paris Agreement (2015) is an agreement between the leaders of developed and developing countries. It involves committing to a low-carbon future to limit temperature change to within a 2°C rise. There is also additional economic support for nations to help the most vulnerable communities adapt to a changing climate.	**Successes:** • Hailed as a landmark success as a 'transparent' and 'inclusive' agreement. • Highlights the seriousness of climate change as a global issue. **Limitations:** • Not all of the agreement is legally binding. • A study published in *Nature* in 2017 outlined that 'no major advanced industrialized country is on track to meet its pledges to control the greenhouse-gas emissions that cause climate change'.

| Individual changes in the household | Changes shown on the diagram below would decrease energy use and thus reduce emissions. | **Successes:**
• When one person makes a sustainability-minded change, often others will follow. |
| | Individuals can reduce, reuse and recycle products so that less refuse is sent to landfill sites. This will lead to a reduction in methane emissions. | **Limitations:**
• Some of these changes require initial investment that some people cannot afford (for example, solar panels of additional loft insulation). |

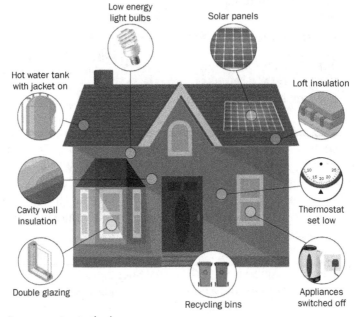

Energy saving in the home

Climate adaptation

Adaptation is the process of changing behaviours to the actual or expected climate and its effects. The Environment Agency states that 'The government is already planning and undertaking specific actions to adapt to current and future climate change.'

Discussion of adaptation strategies:

Positive

1. Some adaptation strategies prevent wider economic implications. The damage from Storm Desmond would have cost £2.8bn rather than £600m without appropriate river defences.

2. Some strategies such as afforestation of river banks also act as habitats for wildlife and act as a carbon sink.

Negative

1. Adaptation strategies do not tackle the root cause of climate change.

2. Some strategies are very expensive and require significant economic investment. This can be a challenge in times of austerity. The Thames Barrier cost over £500m to construct.

3. Adaptation strategies only go so far. The population of the Carteret Islands in the Pacific were the first to be relocated.

ADAPTING TO CLIMATE CHANGE
Examples of strategies

ADAPTING TO DRIER CONDITIONS – INTERNATIONAL

Many agricultural areas are now implementing more sustainable water extraction methods in response to drier conditions. For example, Bill Gates has recently been piloting soil monitoring systems in rural Africa to reduce water use on farms. This has seen water use reduce by as much as 40%.

ADAPTING TO DRIER CONDITIONS – THE UK

The National Adaptation Plan states that the UK government will work to 'restore natural processes' in rivers to buffer against drought and reduce water leaks in infrastructure.

ADAPTING TO RISING SEA LEVELS – INTERNATIONAL

The city of Miami has invested heavily to raise the height of roads and infrastructure to cope with rising sea levels. It is predicted that this could cost $3.2 billion by 2040.

ADAPTING TO RISING SEA LEVELS – UK

Shoreline Management Plans were introduced to provide a strategy for long-term coastal adaptation on a local scale.

ADAPTING TO EXTREME WEATHER – INTERNATIONAL

To protect areas from the intensified tropical cyclones and hurricanes, some governments, such as in Australia, have focused on preserving coastal wetlands, dunes and reefs to absorb storm surges. Structures are also designed to be resilient to high winds and flying debris.

ADAPTING TO EXTREME WEATHER – UK

The UK can expect floods to occur more frequently. The government has invested heavily in local flood defence schemes such as the embankments in Houston, Renfrewshire.

Energy

In this section

- the global distribution of energy resources
- reasons for changes in demand for energy in both developed and developing countries
- the effectiveness of renewable and non-renewable approaches to meeting energy demands and their suitability within different countries

A sufficient and secure energy supply is a significant factor in the welfare and economic development of a society. Meeting the growing demand for energy in a safe and environmentally sustainable way is a major challenge of the 21st century.

The global distribution of energy resources

An energy resource is 'something that can produce heat, power life, move objects, or produce electricity' (Tulane University, 2015). The distribution of the world's energy resources is uneven. Physical conditions and processes vary across countries and continents, and significant investment is required to access and harness these resources.

Coal

- The distribution of coal seams can be explained by past geological processes. Coal is a sedimentary rock that formed over millions of years in wetland ecosystems. China, the USA and Russia have vast reserves.

- Coal seams need to be near the surface so they can be accessed. In Australia, 80% of coal is extracted from open-cut mines.

Coal Proved Reserves, 2015

Total proved coal reserves, measured in tonnes. Proved reserves is generally taken to be those quantities that geological and engineering information indicates with reasonable certainty can be recovered in the future from known reservoirs under existing economic and operating conditions.

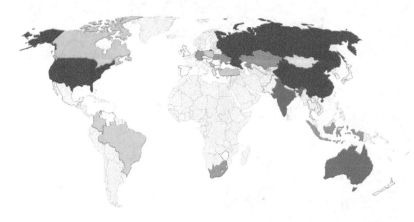

| No data | 0 | 1 billion | 2.5 billion | 5 billion | 10 billion | 25 billion | 50 billion | 100 billion | 250 billion |

Source: BP Statistical Review

OurWorldInData.org/fossil-fuels/ • CC BY

Global distribution of useable coal

Oil

- Found where organic material on the seabed has been converted to oil over millions of years, such as the Middle East.

Wind energy

- Energy can be harnessed from the wind using turbines. Turbines are often found on higher ground or at sea, where they are most exposed to the force of the wind.
- Common in Scotland where winds are strong and consistent. Scotland has 25% of Europe's entire offshore wind power resources.

Geothermal energy

- Most effective near major plate boundaries as these areas have the highest underground temperatures.
- 25% of Iceland's energy is generated by geothermal power plants as the ground surface is thin and close to the Mid-Atlantic ridge.

Hydroelectric power

- Often found in areas of high rainfall, such as in the tropics where convectional rainfall is common (for example, Brazil).
- On major rivers that can be dammed to harness energy (for example, the Yangtze River, China).
- In U-shaped valleys of impermeable rock, which limits water seepage (for example, the Scottish Highlands).

Solar energy

- Photovoltaic solar energy is strongest at lower latitudes and can be harnessed by solar panels.
- In areas dominated by high pressure systems (often the deserts), the air sinks, there is less cloud cover and the sun's rays can more easily reach the surface.

Solar panels to harness photovoltaic energy

Go to Our World in Data to view interactive graphs and charts on energy:
https://ourworldindata.org/renewable-energy

The reasons for changes in demand for energy in both developed and developing countries

Developed nations	Developing nations
The demand for energy has risen due to: Increased vehicle ownership as people have become wealthier (there are now more two-car households), such as in the UK. The development of new, more affordable technology has led to an increase in the demand for electricity (for example, mobile phones, tablets and laptops). In many European countries the proportion of one-person households has more than doubled. More properties have been fitted with central heating and air conditioning units, such as residences in the USA. **Demand for energy is expected to stabilise in the future, as:** Cars will continue to become more fuel efficient, with an increase in hybrid vehicles. Energy efficiency in the residential sector will continue to improve – better insulated homes and more stringent laws for landlords to adhere to when letting out properties. Countries have signed up to the Paris Agreement to reduce greenhouse gas emissions, putting pressure on industries to meet energy saving targets.	There has been a marked increase in energy demand in global manufacturing hubs, such as China. Population growth in developing nations has fuelled demand for energy, e.g. India and Nigeria. An increase in population leads to: • the construction of more homes • the manufacturing of more products, such as televisions and refrigerators, as demand increases • the construction of infrastructure projects. Globalisation has led to greater transportation of goods across the world, increasing reliance on energy resources such as oil. Developing nations, particularly the newly industrialised countries (NICs), have seen an increase in car ownership rates as the wealth of the population increases. Car ownership in India is expected to grow by 775% over the next two decades, further increasing energy demands. Many nations in Africa have not seen as much change in energy demand as other regions of the developing world. This is because many African nations have yet to industrialise, e.g. Somalia.

TOP TIP

Note the difference between recent changes and projections. A projection is a forecast or an estimate.

The effectiveness of renewable approaches to meeting demands of energy and their suitability within different countries

A focus on hydroelectric power

- Hydroelectric power (HEP) is the leading renewable source for electricity generation globally.

- According to the United States Geological Society: 'Energy generated by hydroelectric installations can be injected into the electricity system faster than that of any other energy source'.

- Turbines can be constructed at water storage reservoirs, which means that power can be generated from pre-existing infrastructure.

- The independent production of energy reduces the need for a country to rely on fuel imports, making them more self-sufficient.

- Electricity may be lost in transferring from areas of production to areas of higher demand/population.

Conventional HEP stations

Conventional HEP stations dam the river to create water capacity. During drier months reservoir levels can reduce and, in some cases, water needs to be transferred from other drainage basins. Once dams and reservoirs are built, the facilities can operate at very high efficiencies, which helps to meet energy demand.

Run-of-the-river power stations

These are constructed to accommodate the lowest river flow rate. If the river is in spate (running very fast) potential power production is lost as excess water falls through the spillways. The output of run-of-the-river stations is significantly lower than large-scale hydropower and so may not meet demand. They may struggle to meet energy demands at peak times such as early evening due to a rise in the use of home appliances for evening meals.

Pump-storage dams

For example, the power station at Ben Cruachan pumps water to an upper reservoir at times of low demand but releases it again at times of higher demand to meet energy needs.

Other points to consider

- HEP has additional benefits as it provides water services.

- Dam construction can flood large areas of land. This has social consequences such as displacement of people, e.g. the Three Gorges Dam.

- It can obstruct fish migrations.

A focus on wind energy

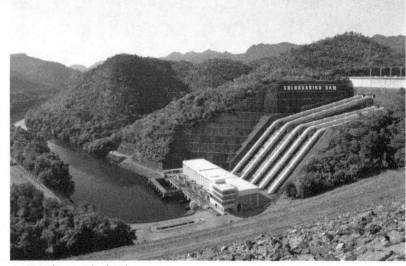

A HEP plant in Thailand

- Wind is an infinite energy resource that is sustainable.

- Analysis from the International Energy Agency in 2019 found that if windfarms were built on all useable offshore sites, they could generate over 36,000 terawatt hours per year. This more than meets the requirements of total global energy demands.

- Wind farms are only a viable development in countries where there is enough land or offshore potential, and a consistent wind supply; for example, in Scotland.

- Output can be variable and may be dependent on the weather conditions.

  ○ When there are periods of no winds or calm weather (for example, when a high-pressure system dominates) the turbines may not turn. The wind farms will then provide very little or no energy. This can be particularly problematic during periods of high pressure in winter, when the temperature is very cold and the demand for energy is greater.

  ○ The wind can be irregular and intermittent.

- Wind energy can struggle to meet energy demands at peak times such as in the early evening due to a rise in the use of home appliances for evening meals. This is because wind energy cannot be stored.

- At times, wind turbines may be switched off due to over-production or dangerously high winds. According the Renewable Energy Foundation, wind farms in Scotland were paid £100 million in 2017 to turn off their turbines.

- Electricity may be lost in transferring from areas of production to areas of higher demand/population.

An offshore wind farm

The effectiveness of non-renewable approaches to meeting demands of energy and their suitability within different countries

A focus on fracking

Fracking is a highly contentious practice. It is a method of drilling used when there are deep deposits of oil and gas underground. There are advantages and disadvantages of fracking, and some of those that refer to the extraction of shale gas are included below:

- The shale gas provides an alternative energy source, reducing reliance on traditional fossil fuels.

- It is a non-renewable energy source so is not a long-term, sustainable solution to meeting energy demand.

- According to one source, the money generated in the US from fracking 'can power our economy and pay for our investments in a smart grid, wind and solar energy, and increased energy efficiency'.

- Fracking is estimated to have offered gas security to the US and Canada for the next 100 years.

- It has presented an opportunity to generate electricity at half the CO_2 emissions of coal.

- Fracking uses vast quantities of water, which is often transported to the fracking site.
  - As some water and its associated additives will be returned to the surface as flowback, chemical contamination may occur. Fracking can contaminate drinking water if it is not done properly.

- The fracking process could be linked to causing tremors in the local area, leading to building damage. Fracking was the 'likely cause' of tremors in Blackpool in 2011.

- Noise and light pollution occur over 24-hour production schedules. There is also an increase in heavy truck traffic, which poses risks to air quality.

- As of 2019, the ban on fracking has been extended in Scotland.

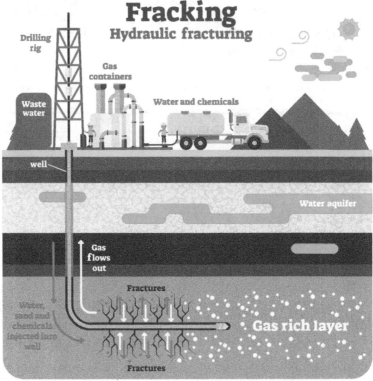

The process of fracking

A focus on coal

- Non-renewable energy provides instant power as and when required, such as meeting demand at peak early evening times.

- In the short term, energy from coal power plants tends to be cheaper than some other renewable energy sources such as solar energy.

- Coal is in abundance in some countries, such as the USA, Russia and China, which reduces costs. However, other countries rely heavily on coal imports, e.g. Japan.

- Coal resources are finite so are not a long-term sustainable solution to meeting energy demand.

- Coal may be considered as ineffective in meeting energy demands of the environmentally-conscious consumer. Many people are now aware of the damaging impact of coal combustion and may choose to use an electricity supplier that only provides energy derived from renewable sources.

Coal combustion and the atmosphere

- The combustion of fossil fuels releases greenhouse gases into the atmosphere. This leads to the enhanced greenhouse effect and climate change.

- The combustion of coal also releases airborne toxins and pollutants into the atmosphere. Some of these substances are linked with acid rain, forest and crop damage. One study has also attributed a quarter of a million premature (early) deaths in China in one year to coal emissions.

Coal power station in Weisweiler, Germany

EXAM QUESTION

Give reasons for changes in energy demand in developing countries.

8 marks

Application of geographical skills

The application of geographical skills section is worth 20 marks and is the second component of Question Paper 2. It consists of a mandatory extended-response question which will require you to apply a range of skills you have developed throughout the course. We have provided two sample questions in this section and given suggestions on how to approach writing extended responses.

Sample question 1

A new outdoor activity centre is to be built in the Cairngorms National Park. The site should meet several requirements:

- There should be enough space for a car park, main activity centre and overnight accommodation for visitors and staff.
- There should be opportunities for zorbing, mountain biking, tree rope courses, kayaking and raft building/racing.
- There should be minimum disruption to local people during the construction process and in the day-to-day running of the centre.
- Coaches should be able to access the site.

Study the Ordnance Survey map extract (Map 1), Source 2 and Source 3 before answering this question. Referring to map evidence from the OS map extract, and information from the other sources:

a) **Discuss** the advantages and disadvantages of the proposed site   **10 marks**

b) **Suggest** any potential social, economic or environmental impacts this development may have on the local area   **10 marks**

Zorbing

Map 1

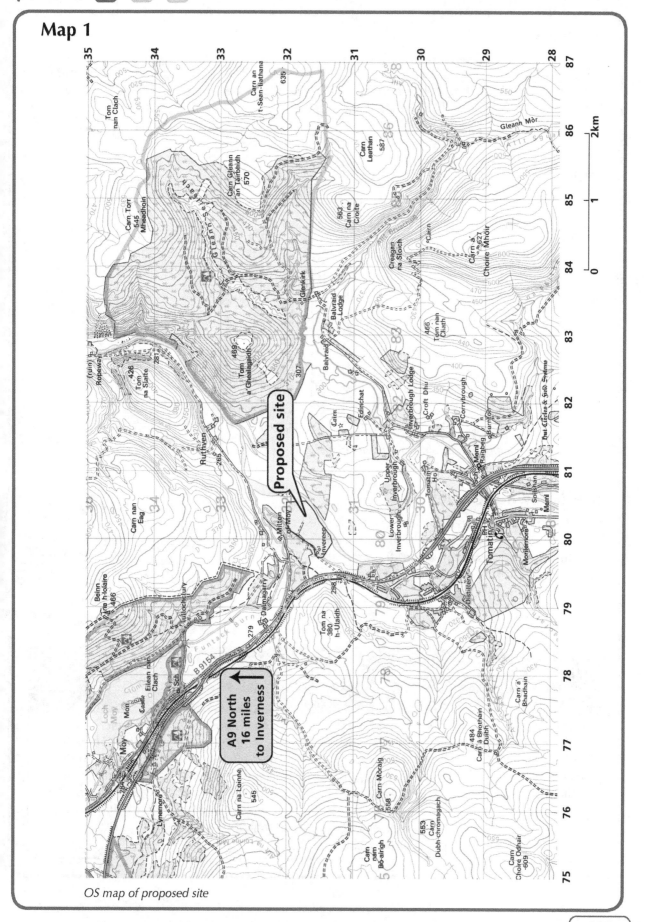

OS map of proposed site

Source 2

'The proposed activity centre is expected to bring over 11,000 visitors to the region each year. The visitors will spend money in the area – boosting the local economy. The activity centre will also provide at least 10 full-time jobs.'

Local MSP

Source 3

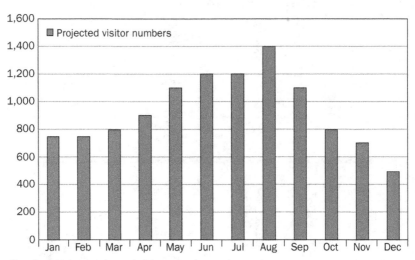

Graph showing projected visitor numbers when the activity centre is in operation

Break down the brief

Before writing your answer, carefully study the brief and sources provided. It is helpful to use the brief when structuring your answer in separate paragraphs. This will help you develop your answer and ensure you hit all necessary points.

TOP TIP

Remember to refer to both the sources and the map. Always incorporate grid references into your response where an OS map is provided.

Text from the brief	Points to consider	Example answer
1. Include a car park, main centre and overnight accommodation for visitors and staff.	• Is the land suitable for building? • What would need to be built for accommodating visitors and staff?	*The land is flat at 8031 which is an advantage for construction. There is also plenty of room to expand the centre if this was required in the future. However, the site is also on the banks of a river (Funtack Burn), which may pose a flood risk.*

Text from the brief	Points to consider	Example answer
2. Provide opportunity for zorbing, mountain biking, tree rope courses, kayaking and raft building/racing.	• Think about the activities and the land they would require. • Is a large area of land needed, should it be flat or steep, etc?	*The site is close to many suitable areas for outdoor activities. It is located next to Funtack Burn which, if conditions permitted, would provide opportunity for some watersports including raft building and racing. The forest in 8031 to the east of the river could be used for the tree rope course. The steep slopes in the north of grid square 8032 would provide conditions for challenging walks/other sports. Kayaking could take place in the local Loch Moy in 7734.*
3. Cause minimum disruption to local residents.	• Look for local buildings and settlements on the map.	*A potential advantage is that the site is 3 km from the nearest village of Tomatin which will minimise disruption to the residents during the construction phase. However, the construction of the activity centre and associated car park could scar the natural landscape. There could also be run off from the building process which could pollute the nearby waterways. The peace and tranquillity of the area could be ruined by large groups of adults and children taking part in the outdoor activities.*
4. Be accessible for coaches.	• Scan the map for local road names.	*The site is located near a main road (A9) providing people access from the cities and Inverness is only 16 miles away.*

When considering the impact of **any** development you may wish to think carefully about the following points:

Positives

- Job creation – construction process.
- Job creation when the development is open – skilled work, stable employment?
- The facilities created by the development may be enjoyed/used by the local population.
- More visitors could be attracted to the wider region.
- Benefits to the local businesses and economy.

Negatives

- Traffic congestion.
- Destruction of wildlife habitats.
- Deforestation – a cause of soil erosion and a potential flood risk.

TOP TIP

Always evidence your points using the source material.

- Pollution:
  - Noise (including construction noise).
  - Air (from construction processes and increased traffic).
  - Water (including run off from the construction of a development, potential flooding).
- Opposition from conservationists.
- Work could be seasonal, and this is not always suitable.

Sample question 2

A well-known property developer has applied for planning permission to the city council for a residential development. Maps 1 and 2 show the proposed site, shaded in purple, at different scales.

The site is predominantly made up of grassed areas that slope slightly towards the south-east corner. There is also a portion of 'brownfield land' upon which the former Simshill Primary School building stood.

The site is easily accessible by public transport, with both King's Park and Cathcart train stations within one mile of Simshill Road. There is currently limited vehicular access to the site; however, there is potential to create a wider access route from either Simshill Road or Old Castle Road.

The site is approximately 2.39 hectares, which is about three football pitches.

Using map evidence and the information from the sources below:

(a) **Discuss** the suitability of the new residential development.

10 marks

(b) **Evaluate** the social, economic and environmental impacts of the new houses.

10 marks

TOP TIP

Make reference to all of the sources, including the OS map, in your discussion of the suitability of the new housing estate. You can focus on positive and/or negative factors.

Source 1

CITY COUNCIL'S POLICY ON BUILDING NEW HOMES:

New homes should:

1. Be environmentally friendly.
2. Make use of sustainable design. For example, orientation of the land for solar panels, possibility of rainwater harvesting.
3. Provide recycling facilities.

New homes must avoid:

1. Loss of biodiversity. Must compensate for any unavoidable loss of habitat by incorporating green features such as including larger gardens and green roofs.
2. Polluting the site and the surrounding area (noise, visual, light, air).
3. Building on a flood risk area.

Source 2

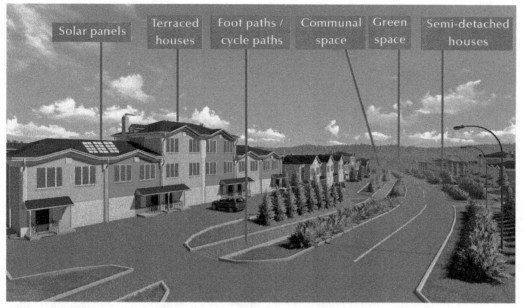

Solar panels | Terraced houses | Foot paths / cycle paths | Communal space | Green space | Semi-detached houses

Architect's plan for the site

Source 3

	Cathcart	Glasgow average	Scottish average
Neighbourhood index			
0 = poor/low		**10 = excellent/high**	
School attainment	9	7	8
Safety (low crime rate)	9	6	8
Average income	8	6	8
Access to public transport	8	9	7
Quality of the roads	8	5	6
Air quality	9	6	8
Range of local shops	7	8	8
Range of leisure facilities	9	6	6
Access to health service	8	6	7
Employment rate	9	6	7
Total	84/100	65/100	73/100
Average property price	£255,813	£160,427	£181,457

Neighbourhood index of Cathcart, compared to Glasgow and Scottish averages

Map 1

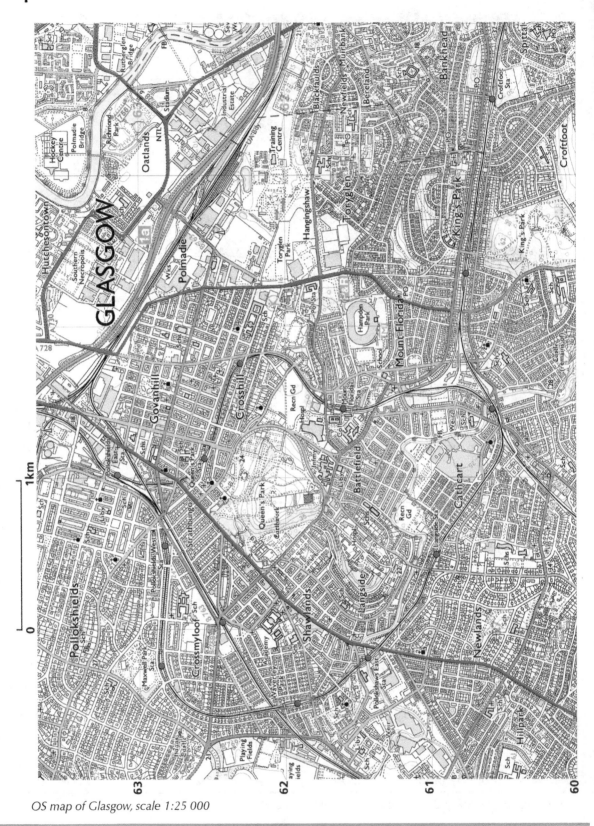

OS map of Glasgow, scale 1:25 000

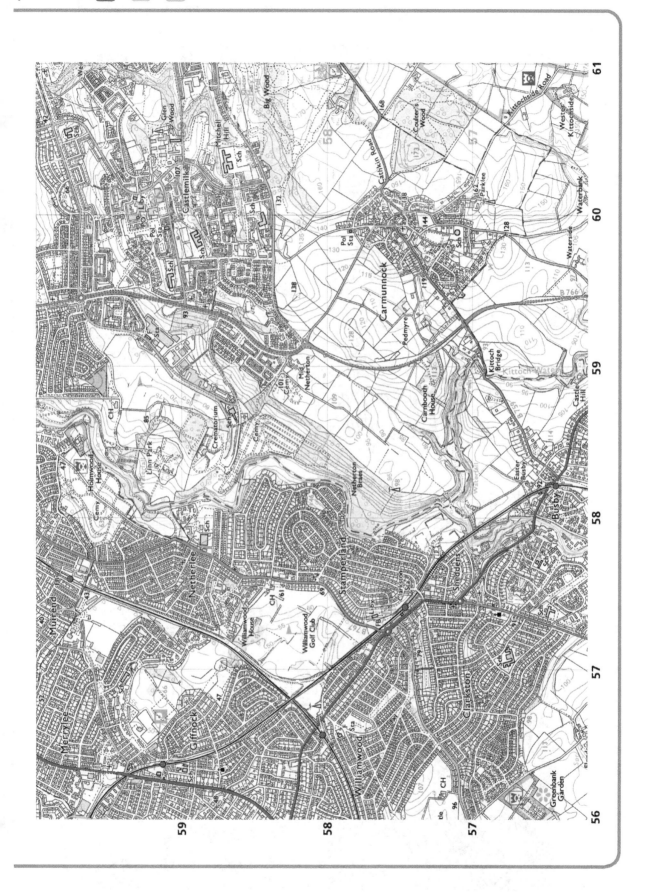

Map 2

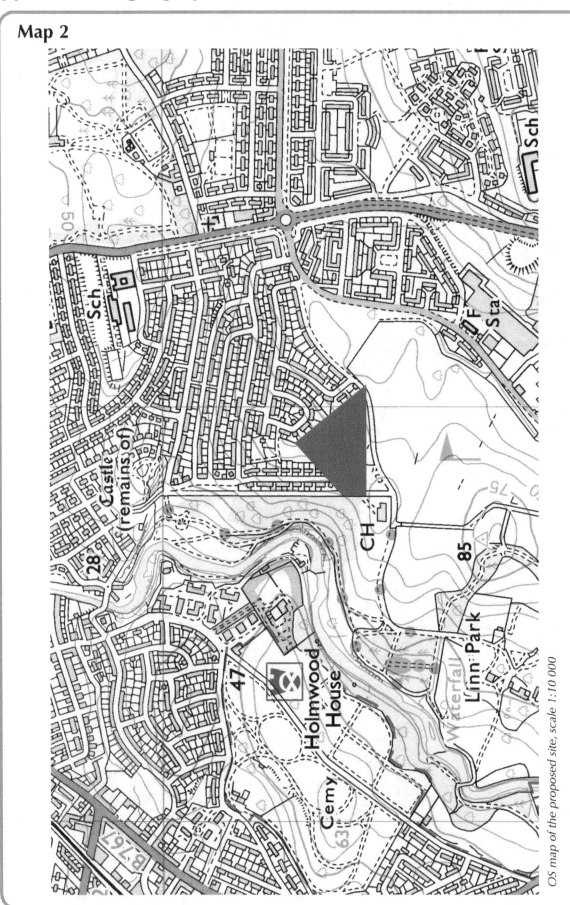

OS map of the proposed site, scale 1:10 000

Writing your answer

Below are some points to consider when answering this question. Look at the maps and sources carefully – can you see any other important factors you should include in your answer?

Part (a)

- The proposed residential development site can be found at GR5859 on Map 1. Looking at this map it is obvious that the site isn't flat. This may add to the cost of the project as it is more difficult to build houses on a slope than on level ground.

- However, as the site is on a slope, any rain water will naturally drain away to the surrounding lower ground. This may mean that a simpler and cheaper drainage system can be planned and built. It is also less likely to flood, fulfilling one of the council's specifications from Source 1.

- The site is located at the edge of the Cathcart suburb. Source 3 shows that this is an affluent neighbourhood with a high average property price (£95,386 more than the Glasgow average) and access to good schools and health services. Map 1 shows that there are at least 10 school buildings (both primary and secondary schools) close to the site.

- Maps 1 and 2 show that the site is well served by roads (many B roads that link to either the A728 or A730) and railways (Cathcart and King's Park) making commuting into Glasgow or East Kilbride easy.

- A significant area of the site is facing south or south-west, which would be an ideal aspect for the installation of solar panels (as shown in Source 2). This will make the house more environmentally friendly than the older houses in the area, fulfilling one of the council's specification from Source 1.

- The site is very close to Linn Park and a golf course (588593 on Map 1). It is also close to a leisure centre (601592 on Map 1) in Castlemilk.

Part (b)

Socio-economic impact:

- The site isn't very big (under 3 hectares) and so it is likely that only a relatively small number of houses will be built. The local population will therefore only increase by a small amount and the current characteristics of the neighbourhood will not be changed – there should be no loss of 'community feeling'.

- The building work will provide jobs.

- Local shops and services are likely to benefit from the slight increase in population.

- The GP surgery and local schools will need to accommodate the new population, but as it is expected to be a relatively small increase, this shouldn't stretch the local services.

- Someone in the local area may object to the planning permission no matter how 'good' the plan, and this may create tension in the local community.

Environmental impact:

- As this site is a former primary school ('brownfield' site) there isn't going to be a great loss of biodiversity or loss of habitat for the local wildlife, fulfilling one of the council's specifications from Source 1.

- The local area may experience a slight rise in noise pollution during the rush hour due to increased numbers of cars in the areas, which is contrary to one of the council's specifications from Source 1.

Glossary

Albedo: The proportion of the incident light or radiation that is reflected by a surface. Forests have a low albedo, whereas ice has a high albedo.

Atmosphere: The layer of gases that surround the Earth consisting mainly of nitrogen and oxygen.

Attrition: Rock fragments (stones and pebbles) hit against each other and so are reduced in size.

Biodiversity: The variety of plant and animal life in the world or in a particular habitat.

Blowhole: This is a crack or fissure in the headland through which air and sea spray is expelled when waves break on the shore.

Census: Population count and survey that is carried out every 10 years in most countries.

Cirrus: Thin, wispy clouds.

Coriolis effect: The anticlockwise rotation of the Earth which deflects winds to the right in the northern hemisphere and to the left in the southern hemisphere.

Counter-urbanisation: When people move out of the city and inner city to live in the countryside.

Cumulonimbus: A dense, towering, vertical cloud associated with thunderstorms and atmospheric instability.

Demographic: Study of population structure.

Differential erosion: When materials erode at different speeds. This is due to one being more resistant than the other.

Emigration: When people move away from one country.

Evapotranspiration: The combined process of evaporation and transpiration.

Forced migration: The movement of people from their country of origin to another in search of safety from war, environmental disaster or religious or ethnic persecution.

Geothermal energy: Thermal energy generated and stored in the Earth.

Glacial isostasy: The rise of land masses that were depressed by the huge weight of ice sheets during the last glacial period.

Glacier: A slow-moving mass of ice.

Greenhouse gas: A gas in an atmosphere that absorbs and emits radiation and therefore contributes to global warming. These gases include carbon dioxide and methane.

Groundwater abstraction: The use of water from underlying rock layers.

Gyre: The circular system of ocean currents rotating clockwise in the northern hemisphere and anticlockwise in the southern hemisphere.

Harmattan: A very dry, dusty easterly or north-easterly wind on the West African coast, occurring from December to February.

Human Development Index: A measure developed by the United Nations Development Programme that ranks national development-based life expectancy, educational attainment, and per capita income.

Hydraulic action: When water, under great pressure, gets into cracks in rock and the pressure squeezes the air, loosening pieces of rock.

Immigrant: A person who comes to live permanently in a foreign country.

Immigration: The process of coming to live permanently in a foreign country.

Insolation: This is the total amount of solar radiation energy received on a given surface area during a given time.

Irrigation: The artificial application of water to farmland.

Isohyets: A line on a map connecting points having the same amount of rainfall in a given period.

Long-term migration: When people decide to settle somewhere else permanently.

Megacity: An urban settlement with a population of more than 10 million.

Migration: The movement of people from one place to another. Movement from within a country is called internal migration, and between two different countries it is called international migration.

Milankovitch cycles: The collective changes in the Earth's movements. It is named after the Serbian geophysicist and astronomer Milutin Milanković.

Ocean acidification: The ongoing decrease in ocean pH – it is caused by the uptake of CO_2 from the atmosphere.

Refugee: Term given to people who flee their home country in search of safety.

Rural–urban migration: When people leave their homes in the countryside in search of education or employment in the city.

Short-term migration: Seasonal migration of agricultural workers or international students studying abroad for their degrees, for example.

Subsistence farming: When farmers grow crops and rear animals to feed just themselves and their family, and not to sell.

Sunspots: Temporary phenomena on the sun – they are dark spots where the magnetic field is stronger than anywhere else on the sun.

Truncated spurs: Before glaciations, valleys are made up of interlocking spurs that rivers meander around and form. A truncated spur forms when a glacier erodes the sides of a V-shaped valley, and cuts off the interlocking spurs. This leaves a steep ridge on the valley sides.

Urban sprawl: When towns and cities grow out into the countryside.

Urbanisation: The growth of towns and cities.

Urban–rural migration: When people leave their city lives in search of a quieter life in the countryside.

Voluntary migration: The movement of people from one place to another based on family links or economic prospects.

Westerlies: The belt of prevailing westerly winds in mid-latitudes in the southern hemisphere.

Specimen answers for sample exam questions

Atmosphere

Sample exam question 1 (page 9)

The tropical regions have an energy surplus, as the amount of solar insolation received at the Earth's surface exceeds that of the re-radiation levels. The Sun's rays are concentrated over a small surface area (Y on the diagram) in the tropical regions due to the more direct, vertical strike of the incoming insolation. Moreover, the equatorial regions receive consistent energy (around 12 hours of daylight each day) throughout the year due to the lack of seasonal tilt, a further contributing factor of an energy surplus. The colour and composition of the Earth's surface also affects the reflection of solar insolation. More insolation is absorbed in the tropical areas where the dark, lush vegetation has a low albedo, thus leading to an energy surplus.

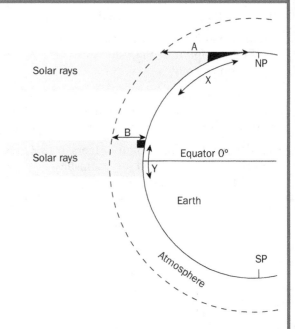

In contrast, the polar regions have an energy deficit as the re-radiation of long-wave energy exceeds that of the incoming solar insolation. Levels of solar energy are low at the poles regions as there is more atmosphere (A on the diagram) for the rays to travel through than at the tropics. This means that more insolation is reflected and absorbed by dust particles, gas molecules and clouds, leading to limited energy being received at the surface. Moreover, the Sun's rays are dispersed over a larger geographical area (X) in the polar regions due to the curvature of the Earth and the lower angle at which the solar energy strikes. As the Earth is tilted on its axis, the poles receive virtually no daylight for six months of the year, which contributes to an energy deficit. Lastly, the albedo effect is more pronounced at the poles where the surface is often light-coloured ice with a high reflective potential.

Sample exam question 2 (page 13)

The sheer size of the Atlantic Ocean allows for the formation of ocean currents. Between 10 and 20 degrees north, the warm North Equatorial current is heated by the intense solar radiation in the tropical regions. It moves west, driven by the prevailing direction of the trade winds.

The Gulf Stream is a warm ocean current that moves north-east across the North Atlantic Ocean. Its movement can be partly attributed to temperature differences between the Gulf of Mexico and the mid-latitudes. The water moves to address an imbalance in density, but it is also aided by the Westerlies, which apply stress to the ocean surface. This frictional drag leads to the formation of a particularly strong and influential current. As the Gulf Stream reaches the European land mass, some of the water is deflected north as the North Atlantic Drift, while some moves southwards, forming part of the cold Canary Current.

In the most northerly regions of the Atlantic Ocean, water is colder due to a lack of incoming solar insolation. The cold, polar water is denser, and thus sinks and moves south to address the imbalance in density. These currents are known as the East Greenland Current and the Labrador Current.

When the currents complete a full loop, they are referred to as 'gyres'. In the North Atlantic Ocean, these gyres move in a clockwise direction due to the Coriolis effect.

Hydrosphere

Sample exam question 1 (page 20)

The drainage basin is an open system, with inputs, outputs, stores and transfers. Solar energy powers the processes within the drainage basin system and might be considered the main input; the other input into this system is precipitation. Water is transferred around the system by surface runoff when the rainfall falls onto an impermeable surface. If the rock is permeable, the rainfall will percolate through the pores in the rock, or it might be infiltrated into the soil. It will then move downwards until it reaches the water table. From there it will move through the ground as throughflow, where it will eventually flow into the river. When the precipitation falls onto plants, it is intercepted and flows down the trunk of the plant as stem flow, where again it will meet the water table. The main outputs in a drainage basin system are evaporation from rivers and lakes as well as transpiration from plants and other vegetation. Water will also be lost from the system when the river flows into the sea. There are many places where water can be stored in the system, which may be on the surface or underground. The main surface storage includes rivers, lakes, glaciers and vegetation, while underground storage includes groundwater and soil storage.

Sample exam question 2 (page 30)

Rainfall began at 05.00 on 26 July. It reached a peak at 08.00 when 6.25 mm of rain fell. The river levels remained steady until 03.00 where they began rising slowly, possibly due to an earlier period of rain. A more rapid increase in discharge occurred at 11.00 when the rising limb increased from 0.5 m to a peak of 0.7 m at 18.00. The lag time therefore was 10 hours. The longer lag time may be due to flat or gently sloping land meaning it takes longer for the rainfall to reach the river. There may also be a lot of soil cover and vegetation in the area, which infiltrates and intercepts the rainfall into the soil, making it take much longer to reach the river channel. It might be quite a large drainage basin, which will take longer for the rainfall to reach the river. From 14.00 to the end of the graph, the river levels continue to drop to 0.5 m at 24.00.

Lithosphere

Sample exam question 1 (page 41)

Where two corries cut back into a mountainside, the intervening ridge is known as an arête.

Corries are formed when snow gathers in a mountain hollow, usually at high altitude where the temperatures are cold enough to allow for the accumulation of snow. Over time, the snow compresses, turning to firn/neve, then ice. This ice then moves downhill as a 'cirque' glacier, its movement aided by gravity.

During the formation of an arête, water enters cracks on the corrie backwalls, freezes and expands (by approximately 9%). This forces the cracks wider and, when repeated regularly, the rock will break apart in a process known as freeze-thaw weathering. This process provides material (scree) for the cirque glaciers to use to erode. Freeze-thaw weather also leads to the jagged appearance of the arête itself.

The cirque glaciers then freeze onto the underlying rock and pluck rock fragments and scree from the surface. This material is then embedded in the sides and base of the glacier. As the glaciers move down the mountainside, the backwalls of both corries become much steeper.

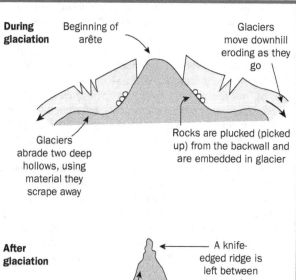

During glaciation — Beginning of arête — Glaciers move downhill eroding as they go — Glaciers abrade two deep hollows, using material they scrape away — Rocks are plucked (picked up) from the backwall and are embedded in glacier

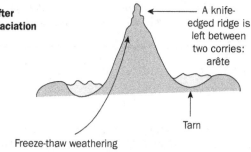

After glaciation — A knife-edged ridge is left between two corries: arête — Tarn — Freeze-thaw weathering

Answers to questions

The base of both hollows become deeper through abrasion – where fragments of rock, embedded in the bottom of the glacier, scrape and wear down the rock underneath (like sandpaper) as the glacier moves. A process known as rotational slip, where the cirque glacier moves back and forward within the hollow, can further deepen the base of the corries.

After the ice melts, two over-deepened hollows are left in the mountainside. The knife-edged ridge between the corries is an arête, an example of which is Striding Edge.

TOP TIP

Remember to include examples.

Biosphere

Sample exam question 1 (page 57)

Podzol soils form in cool, wet climates where precipitation exceeds evaporation. The wet climate leads to leaching and podzolisation, which cause an impermeable iron pan to form in between the A and B horizons. Leaching and podzolisation is when iron and aluminium oxides are removed from the A horizon to the B horizon. The cold climate means that decomposition is very slow and therefore there is only a very thin layer of humus in a podzol soil. Podzol soils are found under coniferous forests. This creates acidic or mor conditions in the soil. Coniferous trees do not shed their leaves every year and therefore there is only a small amount of litter; also, pine needles do not decompose very quickly. There are very few soil biota found in podzol soils due to the acidic conditions and because of the cold climate. This means that podzol soils have very distinct layers. Podzol soils are found in high areas. The formation of the iron pan between the A and B horizons means that the A horizon is often waterlogged as precipitation cannot permeate through the iron pan; this can lead to gleying in the A horizon.

Population

Sample exam question 1 (page 63)

In some developing countries there are a large number of tribal people who migrate in search of better pasture, better farming ground or better opportunities. There are also indigenous people in remote corners of the Brazilian rainforest; these people may be missed out on the census or, due to their mobile nature, counted twice. Countries like South Africa and Kenya have large urban populations who live in shanty towns and in other informal settlements. People living in such accommodation will be difficult to count because they do not have proper addresses to which the census forms may be sent out, or for the enumerator to visit. Countries such as Sudan and D.R. Congo have vast territories to cover with some challenging terrains. In addition, Sudan and D.R. Congo also suffer from poor communication and crumbling infrastructure, which make it extremely difficult to reach everyone, and those living in remote villages are often unaccounted for. In Papua New Guinea over 800 languages are spoken; other countries like Nigeria (500) and India (427) also have the same problem of needing to translate the census into the vast numbers of languages and dialects. It is impossible to accurately translate every word in every question and it is also very difficult to collate all the information onto one useable database. The considerable costs involved in printing, training enumerators, distributing forms and analysing the results can make conducting a census impossible, especially when the country may have more pressing problems like housing and education. In Burkina Faso, 80% of the population is illiterate, which means people require enumerators to complete the census, and this could lead to errors and omission. People may be suspicious and unwilling to give private information to an enumerator and may provide false information. Countries that recently suffered civil war (Syria) and ethnic tension (Rwanda) may find it very difficult to collect certain information such as religion and ethnic group, as people may be afraid of prosecution.

Sample exam question 2 (page 73)

Donor country, e.g. Poland

Advantages

Large numbers of young families have left Poland, so pressure on local services such as education, health care and housing is reduced. The main demographic who immigrate are the working-age population, therefore *pressure on jobs is reduced and levels of unemployment will fall. As young working-age people move away, the birth rate is lowered so population growth rates will slow. Money sent home by the migrants will boost the local economy and migrants will learn new skills and may then return with these skills to their home country.*

Disadvantages

The active and most educated people leave, known as the 'brain drain', resulting in a skills shortage in donor countries. Poland is experiencing a skills shortage in jobs related to building and the service industry. Families are divided and death rates may increase due to the ageing population. Family members remaining in the country of origin may become dependent on remittances being sent home by migrant workers.

Recipient country, e.g. Scotland

Advantages

The short-term gap in labour is filled. Many migrants are highly skilled, e.g. engineers and academics. Migrants will take jobs that locals did not want and will work for lower, more competitive wages, thus reducing labour costs. Migrants will enrich the culture of the area that they move to with language (according to the 2011 Census, 1% of the UK's population speak Polish – which means Polish is the most commonly spoken language after English), food and music – for example, most supermarkets now sell Polish groceries. The increased population will result in an increase in the tax paid to the government, which can be invested in improving local services.

Disadvantages

Migrant workers may feel discriminated against due to feelings of resentment created by fewer jobs being available; this in turn may raise unemployment figures for the local population. Ghettos/slums may develop in parts of cities and there may be a shortage of affordable housing, for example Govanhill in Glasgow. The cost of providing services for the migrant population and their families will increase, e.g. for schooling, health care, etc.

Rural

Sample exam question 1 (page 79)

a) *Stone bunds or lines are placed across the slope parallel with the contours to reduce surface runoff. These lines dam the rainfall, giving it time to infiltrate. The lines can take up some of the available crop land, but they have been shown to increase yields and reduce soil erosion. They are very cost effective as no special equipment or materials are needed, and the building of the lines can be done by members of the community. In Mali and Burkina Faso, this method has been successful and some crop yields have increased by as much as 50%.*

b) *Another strategy that has been effective in the Tigray region of Ethiopia is keeping grazing animals, particularly goats that eat all types of vegetation cover, in an enclosed area rather than letting them wander over the landscape. This means overgrazed ground is kept to a minimum and vegetation cover in other areas remains or has a chance to re-establish itself, leading to a reduction in surface runoff and soil erosion. The animals' feed is supplemented with other fodder that is grown elsewhere. However, fencing can be too expensive for the farmers, or may not be available. In some areas (such as Korr in northern Kenya) whole woodland areas have been destroyed by farmers harvesting material to construct livestock enclosures; these woodlands may have previously been a source of food for the animals. In addition to this, all the farmers in the area may not agree on the need for fencing, which could cause tensions and also make the strategy less effective if some animals in the area were still free to roam.*

Sample exam question 2 (page 79)

As crops fail, families turn to 'famine foods', e.g. leaves, to supplement their diets. Such foods are nutritionally poor, the population's health will suffer and people may starve to death. Rural–urban migration increases as people are 'pushed' to leave the countryside for the cities in search of food and work, and here they will often end up living in informal settlements (shanty towns) on the edge of urban areas. This migration threatens the traditional nomadic way of life and leads to a demographical imbalance as people leave the area. As food and water become scarce, different tribal groups may be forced to live closer together near sources of water, and this can cause ethnic tensions. It can also lead to over-cultivation of that land as people are forced to farm much closer to one another. As income falls, education levels will likely also fall as fewer people with the skills to teach will be left in the area, and the loss of income will mean many families will not be able to afford schooling for their children in any case. International aid is often necessary to ensure the survival of communities, but this can lead to an over-dependence on aid.

Urban

Sample exam question 1 (page 88)

Glasgow

In order to improve the flow of traffic in the city centre, Glasgow city council have introduced one-way systems: St Vincent Street and Hope Street are examples. This solution has somewhat improved the flow of traffic; however, it can be frustrating to drivers who need to drive the 'long way' in order to get to their destination. Drivers who are not used to the one-way system can also find it confusing and therefore drive more slowly, which could cause traffic congestion.

Traffic wardens patrol the streets of the central business district (CBD) in order to discourage drivers from parking illegally. Drivers are fined if they disobey the parking rules, which generates funds for the city council. The council have also made street parking more expensive in order to discourage people. In recent years, a few multi-storey car parks have been built in and around the city. These car parks are well used, especially during the weekend, and this has reduced the amount of cars parked on the street.

Glasgow city council, with joint funding from the Scottish Government and the European Union, have made improvements to old roads, built new ones (bypass) and extended the M74 motorway. Building new infrastructure will help to reduce traffic. However, increased accessibility may encourage more people to drive, so it can also lead to more cars on the road. New road bridges (Millennium Bridge) and footbridges (Tradeston (known as the 'squiggly' bridge)) have been built in recent years to ease the flow of traffic and encourage people to walk to work.

In order to encourage more people to use the public transport system, many improvements have been made. New trains and buses with Wi-Fi have been very popular with commuters. Train and subway stations have been renovated. There are plans to renovate Queen Street Station in Glasgow.

Designated bus lanes throughout the city have improved the flow of traffic. These lanes can only be used by buses, taxis and bicycles; this is an inconvenience to a car user but may encourage more people to take the bus or cycle to work.

There are three Park and Ride subway stations close to Glasgow. This scheme allows drivers to park at the subway station car park at a reduced rate. Subway stations such as Shields Road have the capacity to park 800 cars.

Some firms and companies located in the city have introduced 'flexi-time' – this means commuters do not need to all rush into the city at the same time, but can stagger their arrival and departure time.

Sample exam question 2 (page 90)

Kibera, Kenya

The population of Kenya's largest slum is growing at 5% annually; this means overcrowding is a major problem. The extremely high population density creates many problems as already over-stretched resources such as water and electricity have to be shared more widely, creating tensions within the slum community. The combination of poor nutrition and lack of sanitation accounts for many illnesses and high death rates, leading to higher infant mortality rates and lower life expectancy.

In Kibera, the majority of the population rent their accommodation from private landlords. The rental market isn't regulated and is mostly untaxed by the central government. Private landlords are not obliged to invest money in order to improve the standard of accommodation, therefore people rent small, substandard shacks, without running water or electricity. Family members and often extended family members share the small space. Due to lack of money, the majority of the dwellings are built using cheap and unsuitable building materials, often with no proper ventilation in the kitchen.

Unemployment and underemployment is high in Kibera. Many work in the informal sector doing manual and unskilled jobs, which are often poorly paid. These jobs are also not taxed so the Kenyan government cannot raise tax revenue from the economic activities in the slum. Due to the poor living conditions, lack of jobs and chronic poverty, the area suffers from drug and alcohol abuse and a high crime rate.

The environment quality of Kibera is very poor. Due to the high population density and use of firewood and kerosene for cooking and lighting, the air quality is poor. In addition, the area suffers from rush-hour traffic congestion. There are some organised refuse and recycling collections but not enough to clear the rubbish created by one million people, and as a result the streets and rivers are covered with rubbish. Water quality is also very poor due to constant contamination by sewage and other human waste.

River basin management

Sample exam question 1 (page 96)

Physical factors

The dam should be built at a point where there is hard, impermeable rock such as granite as this will ensure that water does not seep away into the rock. There need to be solid and stable foundations on which the dam wall can be safely constructed. There should be a large, deep valley behind the dam so that a large volume of water can be stored. It should be built at a narrow point in the valley to reduce the length of the dam and therefore the building costs. The area should not be at risk of earthquakes to ensure the dam is not damaged by tremors. The drainage basin should be large with a high drainage density and reliable precipitation or supply of snowmelt.

Human factors

The area should have a low population to avoid the displacement of people and businesses and to keep relocation costs low. If possible, the area should be poor farmland to keep compensation payments to farmers low and avoid destroying valuable, fertile lands. It should not be built in areas of historical, environmental or cultural significance as this could lead to political tensions and costly legal action. There should be demand for water nearby; for example, it could be close to agricultural areas so irrigation and power created by a hydroelectric dam could be utilised.

Sample exam question 2 (page 100)

The Colorado in North America

There are forty million people in seven US States living within the Colorado basin, all needing fresh water. The Colorado River had to be managed because its annual flow was very unpredictable; annual rainfall in this part of the USA is very seasonal and in places very low due to the arid/semi-arid climate (there is less than 250mm of rain annually in the lower basin). However, during the summer months (July/August) violent thunderstorms resulted in flash floods, and a lack of infiltration led to large amounts of surface runoff, causing soil erosion. There were also large amounts of transported sediment, which clogged the river channel, increasing the risk of flood. The lower Colorado basin has a large and growing population, all demanding water for domestic use, farming, recreation, hydroelectric power etc., yet it had a water deficit, whereas the mountainous upper Colorado basin had a water surplus – this needed to be balanced. In addition to this, the USA is a rich and developed nation and wanted to 'green' its desert areas and allow people to settle there – cities of this kind including Las Vegas, Phoenix and Tucson are among the fastest growing cities in the USA.

Development and health

Sample exam question 1 (page 104)

Natural resources/minerals

Oil-rich countries like Saudi Arabia can earn lots of money from selling their oil. Other countries (Brazil, Ecuador) that have lots of mineral reserves will benefit in the same way.

If foreign companies obtain rights to drill and mine for minerals in these places, the government (Brazil, Ecuador) can charge various taxes that will benefit these developing countries. Wealth created by natural resources eventually trickles down to the rest of the population.

On the other hand, countries like Ethiopia have very few natural resources that are in demand, so these countries (also e.g. Somalia, Eritrea) attract very little foreign investment and therefore remain underdeveloped.

Political stability

Countries like Sudan/South Sudan and Syria have unstable government. This means these countries are in the middle of a civil war or a revolution.

Answers to questions

In such places, people are spending all their energy fighting and surviving. Three years of unrest in Syria has set the country back around 35 years (e.g. roads, factories, farms and buildings destroyed).

Countries experiencing civil war will not attract foreign investment as it is too dangerous. On the other hand, countries like India, South Africa and Ghana have good stable government within working democracies. If these places also have natural resources, they are considered to be a great development opportunity.

Strategic locations

Some countries are landlocked and this makes it difficult to trade with other countries. Some countries may also have poor infrastructure and communication so they remain set apart and remote from development (e.g. Afghanistan).

Other countries may be located next to a volatile country that suffers the consequence of poor security (e.g. Northern Pakistan and terrorist attacks).

Countries like Malaysia and South Korea are well located to trade with a wide range of countries. They also benefit from investment from countries like the USA, which continues to have a geo-political interest in the region.

Government attracting foreign companies

Countries like Thailand, Vietnam, South Korea (most Southeast Asia and far Asian countries) have invested a lot of money into educating their population (highly skilled/English speaking) so that they can attract foreign companies to set up their factories.

These countries also offer other incentives (reduced tax, free factory rent sites) to attract investors.

Natural disasters

Bangladesh and Philippines – flood.

Central America – hurricane.

Countries that suffer from natural hazards take a long time to recover and rebuild. This makes them a risky investment and so fewer foreign companies will take the risk.

Tourism

Countries like Brazil (2014 football World Cup, 2016 Olympics) benefit from increased tourism. Growth of tourism brings in jobs that in turn bring valuable foreign currencies ($, £), which in turn improve the living standard of people in Brazil.

Other countries that also benefit from growth in tourism include Morocco and Malaysia.

Corruption

Countries like Zimbabwe and Kenya are underdeveloped due to problems related to corruption. Not all money earmarked for development gets used for that purpose. Some of the money ends up in offshore bank accounts of corrupt government officials. Zimbabwe suffers from hyper-inflation, i.e. the Zimbabwe $ is worth next to nothing on the currency market.

Sample exam question 2 (page 109)

Barefoot doctors

Barefoot doctors have been particularly effective because individuals were chosen by each village to be trained to a basic level of health care and so were fully trusted by the community. In countries with large rural areas it is very difficult to ensure that every village has access to a fully trained doctor/hospital, such as in Uganda.

Oral Rehydration Therapy (ORT)

Oral Rehydration Therapy is the mixture of salt and sugar with clean water to help people suffering from diarrhoea. It is an easy, cheap and effective method of treating dehydration from diarrhoea, allowing even the poorest developing country to tackle this health problem. The World Health Organisation estimates that ORT saves about 1 million babies each year in developing countries.

Organised Health Education Programmes

Non-Governmental Organisations (NGOs) such as Against Malaria educate people about how to prevent diseases spreading, for example, by the use of mosquito nets to prevent malaria. Preventative health care such as vaccinations is easier and more cost-effective than trying to cure someone once they have a disease. The use of village meetings, songs, plays and posters to pass on health education messages are particularly effective in places with an illiterate population, where a written leaflet would be of limited use. Vaccination Programmes to immunise against preventable diseases like polio, cholera, measles, tetanus, etc. are estimated by the World Health Organisation to save between 2 and 3 million lives every year. Charities such as Water Aid work with countries and other aid agencies to improve water and sanitation by installing, for example, pit latrines.

Global climate change

Sample exam question 1 (page 112)

According to records derived from proxy evidence and, more recently, direct measurements, the average temperature of the Earth has fluctuated greatly over time.

One physical driver of climate change is variations in the Earth's orbit and tilt. These variations are known as Milankovitch cycles and are comprised of three dominant cycles: eccentricity, axial tilt and precession. These alter the distance the Earth is away from the sun, leading to fluctuations in incoming solar insolation and, thus, temperature.

Large volcanic eruptions can also cause large quantities of ash and dust to be released into the atmosphere. This ash and dust can then absorb and scatter insolation, which temporarily lowers temperatures (as witnessed after the eruption of Mount Pinatubo). Moreover, volcanoes also release sulfur dioxide, which ultimately creates fine sulfate aerosols in the stratosphere. The reflection of solar radiation is further increased, reducing temperatures.

Another physical factor that has been attributed to climate change is the amount of solar activity (sunspots). Solar activity tends to occur in cycles of 11 years, and when the sun is more active it releases more energy. However, the influence of sunspots on global climate is regarded as minimal in the scientific community.

Moreover, melting ice caps can expose the land underneath, which reduces the albedo (reflectivity) of the Earth's surface. This can result in an increase in temperatures and is known as a positive feedback loop. Melting ice caps can also decrease water temperatures and salinity in the surrounding water. This can alter currents such as the North Atlantic Drift, which accounts for nearly one-third of Western Europe's heat.

Sample exam question 2 (page 115)

Global warming has and will continue to have global ramifications.

An increase in global temperatures will result in glacial retreat, a reduction in land ice cap cover and thermal expansion of the oceans, leading to an increase in global sea levels. Low-lying coastal countries, such as Bangladesh, will be affected most severely by rising sea levels; leading to a large-scale displacement of people, loss of land for farming and the destruction of property. Sea level rise also poses an existential threat to low-lying island nations such as the Maldives (which has an average altitude of only 1 m above sea level).

There will also be an increase in intensity of meteorological events such as hurricanes and droughts. Hurricanes can cause deaths, infrastructure damage and large-scale displacement of people and this will affect people in disaster risk zones such as the Philippines. With some areas becoming even drier, potential conflicts over water supplies may arise, which could lead to war (the Sahel region will be particularly vulnerable to this).

Ecosystems will continue to change. Sensitive coral will continue to suffer bleaching as the ocean warms further, e.g. the Great Barrier Reef off the coast of Australia. This will lead to a reduction in biodiversity and tourism revenue. There is also the potential for the extinction or reduction in the number of certain species (a 60% loss of the Adélie penguin population in Antarctica has been predicted). More people will be at risk from vector-borne diseases such as malaria as the Anopheles mosquito will move into new areas made habitable by changing temperatures and precipitation levels (such as Europe).

Energy

Sample exam question 1 (page 127)

There has been a marked increase in energy demand in developing nations. One reason is because many manufacturing centres have relocated from developed to developing countries, such as China. As the global population has increased (now nearing 8 billion people) and got richer, there is a greater demand for manufactured goods. These products, including refrigerators and televisions, are often mass-produced in the energy-intensive factories of the developing world.

Moreover, the population of developing nations has increased rapidly as the fertility rate remains well above replacement level in many countries, such as Indonesia. More homes and infrastructure need to be constructed to meet the demands of the rising population. The construction process uses vast amounts of energy, but so does the heating/cooling and lighting of these new properties.

Further to this, developing nations, particularly the NICs, have seen an increase in car ownership rates as people have become wealthier. Car ownership in India has risen dramatically and is expected to grow by 775% over the next two decades.

Some nations in Africa have not seen as much change in energy demand as other regions of the developing world due to lower levels of industrialisation, e.g. Somalia.

Higher
GEOGRAPHY

Practice Papers

Kenneth Taylor

Revision advice

The exam

You will sit two final exams for Higher Geography*.

The first exam, **Physical and Human Environments,** lasts 1 hour and 50 minutes and is worth 100 marks. The exam is made up of two sections.

- Section 1: Physical Environments, which is worth 50 marks. **You must answer every question in this section.**

- Section 2: Human Environments, which is worth 50 marks. **You must answer every question in this section.**

The second exam, **Global Issues and Geographical Skills,** lasts 1 hour and 10 minutes and is worth 60 marks.

- Section 1: Global Issues, which is worth 40 marks. There are four questions in this section. You will have studied at least two of these topics. All of the questions are worth 20 marks. **You must answer two of these questions.**

- Section 2: Application of Geographical Skills, which is worth 20 marks. In this section you will be tested on your mapping skills and use of numerical and graphical information. For a good mark you must use the OS map and all of the information and resources given in the question paper. **You must answer this question.**

*Note, the new exam structure was introduced in 2019, so if you look at past papers online prior to this, the structure will be different.

Examination advice

The obvious advice is really important and applies to all of your exams:

- Make sure you know when your final exam is.

- Arrive early for the exam.

- Bring blue or black ink pens and pencils for diagrams.

- Write legibly so that your answers are understood. Consider leaving lines between paragraphs and a short section between each question, so that you have space should you wish to return to this question.

- Make sure you are well aware of the exam structure: how many questions you should answer in each section; how long the exam lasts.

The following apply to the answers you give. These tips are crucial as you may be losing marks without realising it. If you are not sure what any of them mean, look at the tips in the marking instructions section online at www.leckiescotland.co.uk for more detail. Alternatively, ask your teacher and/or classmates.

- Read each question very carefully. Identify the command words (see the next section) and respond accordingly. Make sure you are totally clear what you are being asked before you start writing.

- Plan your answer by writing down key words. Add reminders of important things to include in your answer and cross these off as you go. Do not make this too long, but spending a short time thinking out your answer will definitely pay off.

- In longer questions and/or questions where you are asked to do two things (e.g. social and environmental impacts), use subheadings and paragraphs to provide structure. This makes your answers easier to follow and will ensure that you cover each area of the question (there are penalties if you do not).

- Give full explanations. In Higher, you must explain and give reasons. For example, when covering physical processes such as abrasion and hydraulic action, you must explain in full how these processes operate. Linking words and phrases such as 'this means that', 'therefore' and 'because' will force you to conclude points. Make sure you link these points to the question.

- Watch your time. You must finish both exams to do well. Pace yourself and move through each section or you will not complete the exam. As a rule of thumb, you have a little over 1 minute per mark (i.e. a 10-mark question should take you about 11 minutes). Make sure you look at the number of marks available and vary the length of your answer accordingly.

- Where appropriate, give examples from the case studies you have learned. Remembering place names and named management strategies will be credited where it is asked for in the question and will be expected of very good candidates. Be careful though – learn too many and you will struggle to include them and may get confused.

- Use geographic terminology throughout your answers. For example, when areas of the inner city are redeveloped, this leads to growth as new companies come into the area, creating jobs and improving services, and better candidates will refer to this process as the 'multiplier effect'.

- Practice drawing annotated diagrams that you can reproduce quickly in the exam. Marks will not be given for your artistic ability but for the information that you add as labels.

There are a number of common errors to avoid during the exam:

- Answering more than you need to. Do not answer more than two questions in Section 1 of the second exam. If a question has an option, only write about one option. Similarly, if you write more for a 8-mark question than a 12-mark question, you are going to lose marks somewhere.

- Not answering the whole question. If you are asked for solutions and the effectiveness of these, you will forfeit some marks if you do not talk about both parts of the question.

- Not referring to sources in the Application of Geographical Skills question. You will need to make good use of the map and any numerical data given to you.

- Not completing the exam. Use this book to practice your timing. Your best chance of doing well is to pace yourself evenly through the paper and to complete every question.

- Giving irrelevant information in answers. You must stick to the question. For example, if you are asked about the impact of climate change, you should not write about the causes.

- Writing vague and/or over-generalised answers. In these types of answers, candidates may name a city or country but other than this it appears that they know very little about their case study.

- Reversals. This is where you explain the exact opposite in your answer. This will not gain additional credit. For example, if you were asked about migration you may write about push factors such as lack of jobs and poor health-care. You must be careful that when discussing pull factors you simply do not just write about the opposite, e.g. employment opportunities and lots of doctors in hospitals.

Command words

In the exam, a number of command words will appear. These will appear in **bold** within the question to draw your attention to them. They indicate how you should approach each question. Take your time to reflect on the table below and look at the examples within this book and the corresponding marking instructions.

Each of the command words requires you to answer in a slightly different way. It is important you respond accordingly. For example, if you write in a descriptive way in an explain question, you will limit the number of marks you can achieve. Similarly, if you begin to evaluate or comment on the effectiveness of management strategies but have only been asked to explain these, you will not get credit for the non-relevant sections of your answer.

Command word	Meaning/Explanation
Describe	In this question you are being asked to make relevant factual points. In some situations you may be asked to **describe, in detail**. This requires that you give more information in your answer. At other times you may be given a source such as a graph or table of statistics and be asked to **describe** changes, trends or patterns. In this instance make sure you: • Refer to the units in both the x and y axis in your answer, giving data and making calculations based on key changes. • Use descriptive language to indicate rate of change such as rapid, gradual or steady when referring to increases, decreases or periods of stability on graphs. • In topics such as hydrosphere, there are key terms the examiner would expect to appear in answers; for example, rising limb, lag time etc. Note, you may also be asked to **compare** data sets/more than one graph. Since 2019, **describe** questions have been reintroduced to the exam. They often appear in two-part questions with an explanation of trends or management strategies in the follow-up question.
Explain/suggest reasons	Think about why something has happened. What are the reasons and/or processes behind an action or outcome? For example, if you are asked to **explain** the formation of a corrie, it is important that you are able to demonstrate an understanding of the processes and conditions involved. You will need to give more detail than a describe question to achieve marks. If you limit your answer to descriptive points you will not be able to get more than half marks for that question.
These command words are designed to test higher-order skills.	
Analyse	In this type of question you should give an account of the relationship between factors. Links will be important. For example, you may be asked to **analyse** the various properties in the formation of a soil.

Evaluate	In this question you are being asked if a solution/management strategy etc. has been a success or a failure. You should briefly explain the strategy before assessing the impact it has had. In this question type it is important to back up your answer with evidence and to have a variety of ideas.
Account for	Give reasons – this may often come from a source given in the exam paper. For example, you may be asked to **account for** the rise in global temperatures with reference to physical and human factors.
Discuss	Give the key features of different viewpoints or consider the impact of change. For example, you may be asked to **discuss** the effects of a water-related disease on people.
To what extent/ Comment on the effectiveness of	Similar to 'evaluate' in certain contexts. Comment, with evidence, on the impact of a strategy. For example, **to what extent** have the methods used to try and control the spread of malaria been effective?

Practice Papers for SQA Exams

HIGHER GEOGRAPHY
Physical and Human Environments
Exam A

Duration – 1 hour and 50 minutes

Total marks – 100

SECTION 1 – PHYSICAL ENVIRONMENTS – 50 marks

Attempt ALL questions.

SECTION 2 – HUMAN ENVIRONMENTS – 50 marks

Attempt ALL questions.

You will receive credit for appropriately labelled sketch maps and diagrams.

Write your answers clearly in the answer booklet provided. In the answer booklet you must clearly identify the question number you are attempting.

Use **blue** or **black** ink.

Note: The reference maps and diagrams in this paper have been printed in black ink only. No other colours have been used.

SECTION 1: PHYSICAL ENVIRONMENTS – 50 marks

Attempt ALL questions

Question 1

Diagram Q1: Landscape of glacial deposition

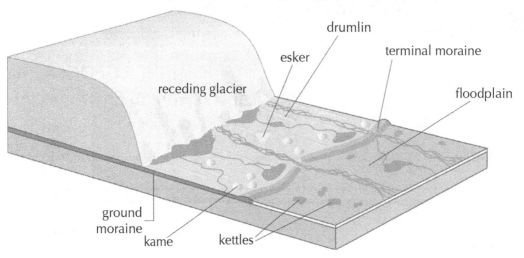

Look at Diagram Q1.

1. **Explain** the formation of **one** of the following features of glacial deposition.

   • Drumlin

   • Esker

   • Terminal moraine

You may wish to use an annotated diagram or diagrams in your answer.   **8**

Question 2

Diagram Q2: Headland and bay

Look at Diagram Q2.

Explain the formation of a headland and bay.

You may wish to use an annotated diagram or diagrams in your answer.   **8**

Question 3

Diagram Q3A: West Africa

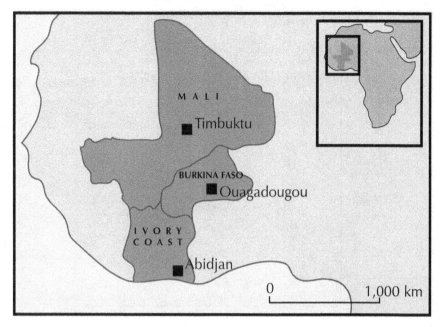

Diagram Q3B: West Africa – selected rainfall graphs

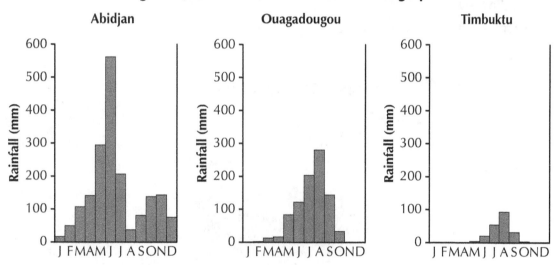

Study Diagram Q3A and Diagram Q3B.

(a) **Describe** the changing rainfall patterns experienced when moving from Abidjan to Timbuktu and

(b) **Suggest reasons** for these changes.

10

Question 4

Explain why there is a deficit of solar energy at the poles and a surplus of solar energy at the Equator.

You may wish to use an annotated diagram or diagrams in your answer   **8**

Question 5

Diagram Q5: Flood hydrographs

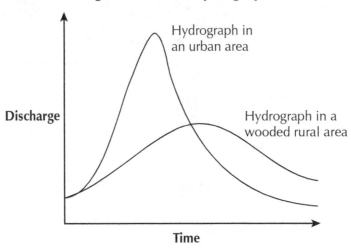

Look at Diagram Q5.

Account for the differences in discharge between the urban and rural hydrographs.   **8**

Question 6

Diagram Q6: Gley soil profile

Look at Diagram Q6.

Explain how a gley soil is formed.

You may wish to refer to conditions and processes such as climate, drainage, natural vegetation, rock type and organic matter.   **8**

SECTION 2: HUMAN ENVIRONMENTS – 50 marks

Attempt ALL questions

Question 7

Diagram Q7A: Population pyramid for Chad, 2010

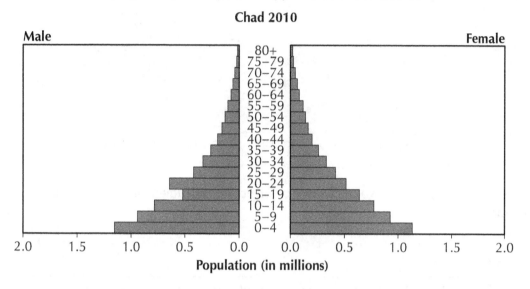

Diagram Q7B: Projected population pyramid for Chad, 2050

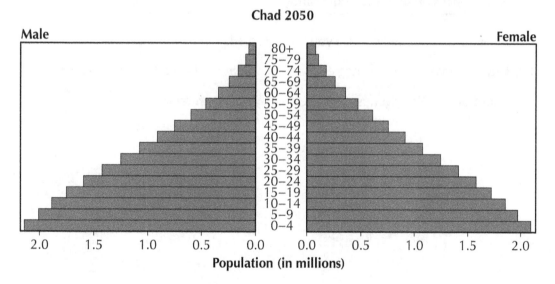

Study Diagram Q7A and Diagram Q7B.

(a) **Describe** the projected changes in Chad's population **and**

(b) **Discuss** the possible consequences of the 2050 population structure for Chad.

You should refer to the impact on the economy of Chad **and** the impact on the people of Chad.

12

Question 8

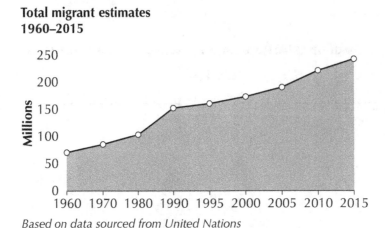

Diagram Q8: Total migrant estimates

Total migrant estimates
1960–2015

With reference to a forced **or** a voluntary migration you have studied, **explain** the causes of this migration.

8

Question 9

Many rural areas, including rainforests and semi-arid regions of the world, are facing the serious consequences of rural land degradation

For a named rainforest **or** semi-arid area you have studied:

(a) **Explain** the management strategies used to combat rural land degradation **and**

(b) **Comment** on the effectiveness of these management strategies.

12

Question 10

Diagram Q10: Traffic congestion in the UK

Look at Diagram **Q10**.

Referring to a city you have studied in the **developed** world:

(a) **Discuss** the main traffic problems facing residents and local authorities **and**

(b) **Explain** the management strategies used to combat traffic congestion.

10

Question 11

Diagram Q11: A shanty town in Cape Town, South Africa

With reference to a named city you have studied in a **developing** country, **explain** the strategies used to manage housing problems in shanty towns.

8

[END OF QUESTION PAPER]

Practice Papers for SQA Exams

HIGHER GEOGRAPHY
Global Issues and Geographical Skills
Exam A

Duration – 1 hour and 10 minutes

Total marks – 60

SECTION 1 – GLOBAL ISSUES – 40 marks

Attempt **TWO** questions.

SECTION 2 – APPLICATION OF GEOGRAPHICAL SKILLS – 20 marks

Attempt the question.

You will receive credit for appropriately labelled sketch maps and diagrams.

Write your answers clearly in the answer booklet provided. In the answer booklet you must clearly identify the question number you are attempting.

Use **blue** or **black** ink.

✕Leckie
the education publisher
for Scotland

SECTION 1: GLOBAL ISSUES – 40 marks

Attempt TWO questions

Question 1 – River basin management

Diagram Q1 River basin physical characteristics

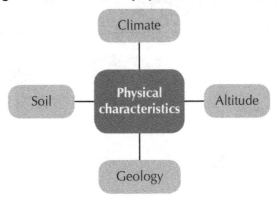

Look at Diagram Q1.

For a named river basin you have studied:

(a) **Analyse** the physical characteristics of the river basin. **8**

(b) **Discuss** the socio-economic **and** environmental impacts of the named river management project you have studied. **12**

Question 2 – Development and health

> 'Resources need to be targeted at improving Primary Health Care if we are ever going to improve the health of people in developing countries.' Aid Worker

(a) Referring to named examples, **explain** why improving health standards through primary health-care strategies is suited to people living in **developing** countries. **10**

(b) Look at Diagram Q2.

Diagram Q2

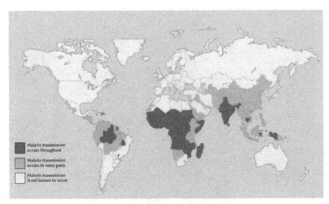

For malaria, or any other water-related disease you have studied, **explain** the human **and** environmental conditions that put people at risk of contracting the disease. **10**

Question 3 – Global climate change

(a) **Explain** the human factors that contribute to climate change. **8**

(b) **Discuss** the impact of global warming on named locations throughout the world. **12**

Question 4 – Energy

(a) Referring to both renewable **and** non-renewable energy sources, **explain** the global distribution of energy resources. **10**

(b) **Comment** on the effectiveness of non-renewable approaches to meeting energy demands. You should refer to a country or countries you have studied in your answer. **10**

[END OF QUESTION PAPER]

SECTION 2: APPLICATION OF GEOGRAPHICAL SKILLS – 20 marks

Attempt the question

Question 5

West Lulworth is a small village on the south coast of England. As well as being popular with tourists, it is a rural community surrounded by many working farms.

The planning proposal below outlines plans for a new holiday park to be constructed in West Lulworth. The plans have been approved but a decision has to be made about which site to locate the holiday park on in the area. Two sites in the area are being considered.

Planning proposal

A plan has been approved in principle for the construction of a holiday park. The park will contain:

a. A camp site with space for 20 pitches (tents)

b. Berths for 10 caravans

c. A shower and laundry block

d. A café and bar

e. A car park for 10 vehicles (parking will also be available next to each caravan)

The investment of £350,000 will be part funded by the regional council, the rural development fund and private holiday firm Buzz Holidays.

Study Diagram Q5A (OS Map Extract of Dorset Coastline on facing page), Diagram Q5B, Diagram Q5C, Diagram Q5D, Diagram Q5E, Diagram Q5F and Diagram Q5G.

Referring to map evidence and other information from the sources:

(a) **Suggest reasons** for the large number of visitors who are drawn to West Lulworth each year. **5**

(b) Decide which site, A or B, is the best site for the proposed holiday park. **15**

You should:

(i) **Discuss** the suitability of the site **and**

(ii) **Suggest** any possible social, economic or environmental impacts this development may have on the area

Diagram Q5B: Dorset coastline

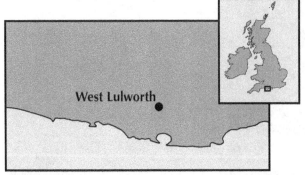

Bristol Airport to West Lulworth – 76 miles

Southampton Airport to West Lulworth – 54 miles

London City Airport to West Lulworth – 127 miles

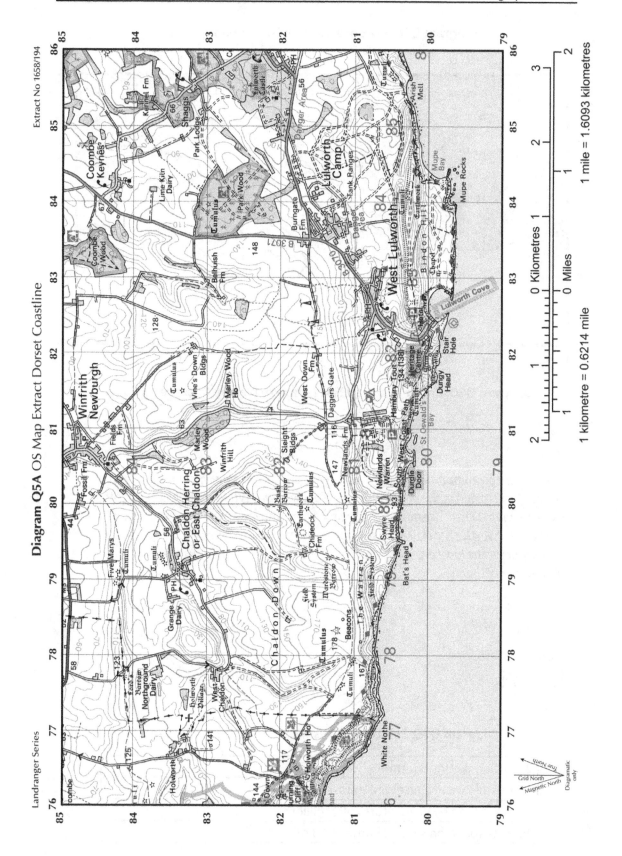

Diagram Q5A OS Map Extract Dorset Coastline

Landranger Series

Extract No 1658/194

Diagram Q5C: Proposed sites of new holiday park

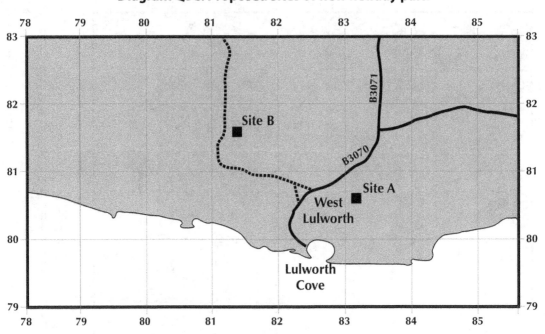

Diagram Q5D: Letter extract, *West Lulworth Weekly Observer*, 6 June

Dear Editor

The lack of accommodation in the local area is putting visitors off coming to West Lulworth or staying in the area longer than a few hours. While young people in other small towns in Dorset are moving into tourist jobs and staying in the area, youngsters in the West Lulworth district are leaving home and moving to nearby towns and cities. It is important for the local economy that the plans for the new holiday park are approved as a matter of urgency.

Yours sincerely

Lisa Hotchkiss

Managing Director

Buzz Holidays

Diagram Q5E: Selected tourism data for Dorset coastline

21 million day visits annually
12% of all jobs in Dorset are tourism related
Annual spend by day visitors £682 million
Tourists spend on second homes £3.3 million

Diagram Q5F: Land use in the county of Dorset

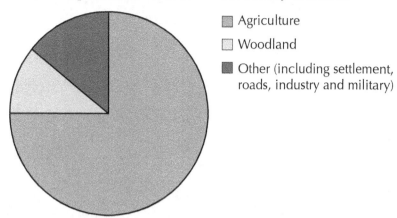

- Agriculture
- Woodland
- Other (including settlement, roads, industry and military)

Diagram Q5G: Employment by broad industrial sector

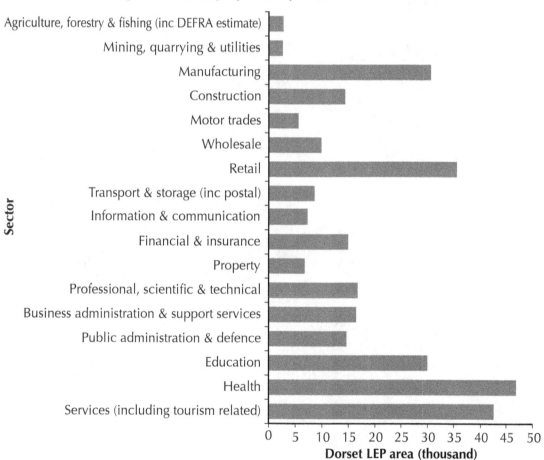

Sector (y-axis)

Dorset LEP area (thousand) (x-axis)

[END OF QUESTION PAPER]

Practice Papers for SQA Exams

HIGHER GEOGRAPHY
Physical and Human Environments
Exam B

Duration – 1 hour and 50 minutes

Total marks – 100

SECTION 1 – PHYSICAL ENVIRONMENTS – 50 marks

Attempt ALL questions.

SECTION 2 – HUMAN ENVIRONMENTS – 50 marks

Attempt ALL questions.

You will receive credit for appropriately labelled sketch maps and diagrams.

Write your answers clearly in the answer booklet provided. In the answer booklet you must clearly identify the question number you are attempting.

Use **blue** or **black** ink.

Note: The reference maps and diagrams in this paper have been printed in black ink only. No other colours have been used.

SECTION 1: PHYSICAL ENVIRONMENTS – 50 marks

Attempt ALL questions

Question 1

Diagram Q1: Pyramidal peak

Look at Diagram Q1.

Explain the formation of a pyramidal peak.

You may wish to use an annotated diagram or diagrams in your answer.   8

Question 2

Diagram Q2: Selected soil profiles

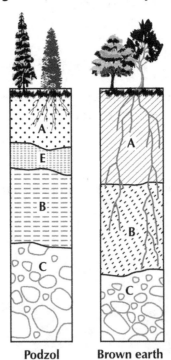

Podzol   Brown earth

Look at Diagram Q2.

Account for the differences in the properties of a podzol **and** brown earth soil.   10

Question 3

Diagram Q3A: Global heat budget

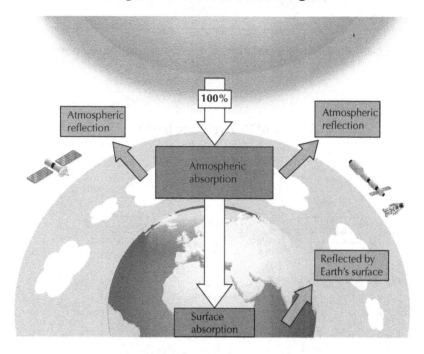

Diagram Q3B: Reflection and absorption of solar energy (%)

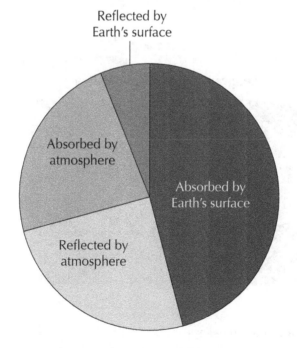

(a) **Describe** the energy exchanges that occur in the atmosphere and at the Earth's surface **and**

(b) **Explain** why less than 50% of the energy arriving at the edge of the atmosphere is absorbed by the surface of the Earth.

10

Question 4

'Energy is transferred from areas of surplus to areas of deficit. Atmospheric circulation helps to redistribute this energy.'

Explain how atmospheric circulation redistributes energy around the globe. **7**

Question 5

Diagram Q5: Drainage basin

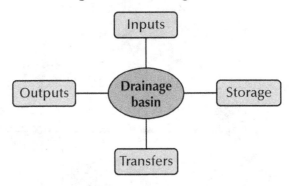

A drainage basin is an open system with four elements.

Discuss the movement of water within a drainage basin with reference to the four main elements in Diagram Q5. **7**

Question 6

Diagram Q6: V-shaped valley

Look at Diagram Q6.

Explain the formation of a V-shaped valley.

You may wish to use an annotated diagram or diagrams in your answer. **8**

SECTION 2: HUMAN ENVIRONMENTS – 50 marks

Attempt ALL questions

Question 7

Diagram Q7: Change in vehicle by type – Maharashtra State, India 2010–2015

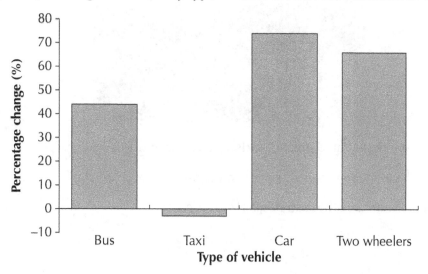

Study Diagram Q7. Mumbai is the capital and largest city in Maharashtra State, India.

(a) (i) **Describe** the changes that have taken place in vehicles by type **and**

(ii) **Discuss** the negative consequences facing cities like Mumbai as a result of traffic congestion.   **8**

(b) For a named city you have studied in the **developing** world **explain** the strategies used to manage traffic-related problems.   **8**

Question 8

Cities in the developed world face many challenges at the start of the 21ˢᵗ century. Many of these problems are a result of changes and failed management in the preceding 60 years. One of the most notable is the provision of housing to meet the needs and living standards of their citizens.

Referring to a named city you have studied in the developed world, **explain** the strategies used to solve housing problems.   **8**

Question 9

Diagram Q9: Lulworth Cove

For a named coastal **or** glaciated upland landscape you have studied:

    (a) **Discuss** the management strategies used to manage land use conflicts **and**

    (b) **To what extent** have these strategies been successful?　　　　　　　**10**

Question 10

With reference to a **voluntary** migration you have studied, **discuss** the impact on the host **and** the donor countries.　　　**8**

Question 11

With reference to countries you have studied, **explain** the challenges faced by **developing** countries in collecting accurate census data.　　　**8**

[END OF QUESTION PAPER]

Practice Papers for SQA Exams

HIGHER GEOGRAPHY
Global Issues and Geographical Skills
Exam B

Duration – 1 hour and 10 minutes

Total marks – 60

SECTION 1 – GLOBAL ISSUES – 40 marks

Attempt **TWO** questions.

SECTION 2 – APPLICATION OF GEOGRAPHICAL SKILLS – 20 marks

Attempt the question.

You will receive credit for appropriately labelled sketch maps and diagrams.

Write your answers clearly in the answer booklet provided. In the answer booklet you must clearly identify the question number you are attempting.

Use **blue** or **black** ink.

×Leckie
the education publisher
for Scotland

SECTION 1: GLOBAL ISSUES – 40 marks

Attempt TWO questions

Question 1 – River basin management

Diagram Q1A: Map of River Nile

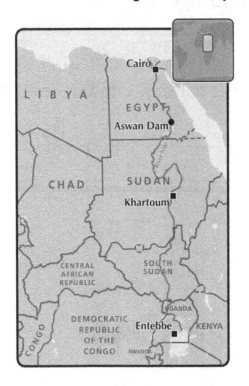

- ■ Sites of climate graphs (Diagram Q1C)

- ● Aswan Dam Water Management Project

Diagram Q1B: Projected population growth in Egypt

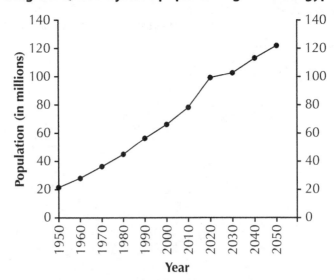

Diagram Q1C: Climate graphs

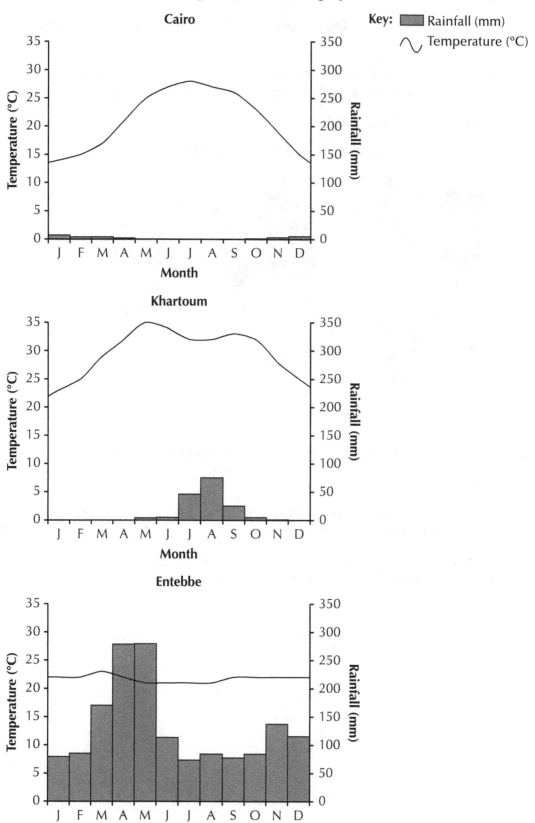

Look at Diagrams Q1A, Q1B and Q1C.

(a) Referring to Egypt and using the resources provided, **explain** why there is a need for water management. **10**

(b) **Discuss** the social, economic and environmental benefits **and** adverse consequences of a named river management project you have studied. **10**

Question 2 – Development and health

Diagram Q2: Human development index

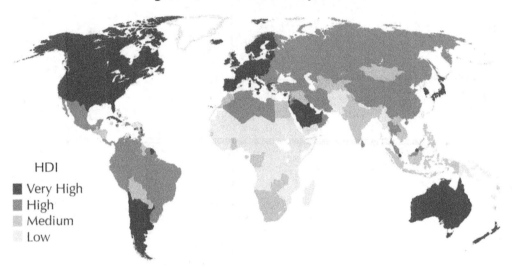

The Human Development Index (HDI) is a composite indicator of development.

(a) Look at Diagram Q2.

Referring to the HDI, or any other composite indicator you have studied, **explain** why it is a useful indicator of development. **8**

(b) **Suggest reasons** for the differences in levels of development between **developing** countries. You should refer to named examples in your answer. **12**

Question 3 – Global climate change

(a) With reference to areas you have studied, **discuss** the impact climate change is having. **10**

During the last 20 years, a range of strategies have been adopted that are designed to slow down rates of climate change and minimise the impact of a changing climate.

(b) **Comment** on the effectiveness of strategies aimed at reducing greenhouse gas emissions **and** managing the effects of climate change **10**

Question 4 – Energy

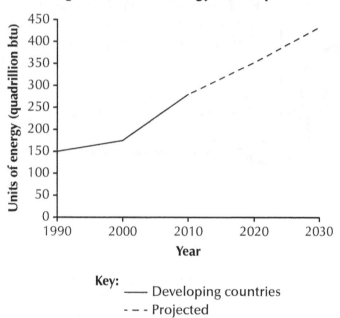

Diagram Q4: World energy consumption

Key:
—— Developing countries
- - - Projected

(a) Look at Diagram Q4.

   (i) **Describe** the changes shown in Diagram Q4.

   (ii) Referring to **developing** countries, **suggest reasons** for the change in energy consumption. **10**

(b) **Comment** on the effectiveness of renewable approaches in meeting energy demands. You should refer to a country or countries you have studied. **10**

[END OF QUESTION PAPER]

SECTION 2: APPLICATION OF GEOGRAPHICAL SKILLS – 20 marks

Attempt the question

Question 5

Running club, Dunoon Hill Runners, have applied for permission to organise a 21-mile trail race on the outskirts of the town. As well as the full distance, the race will also have the option of a team event with two competitors running 10.5 miles each to complete the distance.

Race Briefing – Dunoon Hill Runners 21-mile trail race

Club secretary, Scout Taylor said:

> The route has the potential to become one of Scotland's classic trail races as well as boosting the local economy. Our requirements are for a route that:
>
> - is off road, i.e. not on A or B class roads
> - provides a challenging/hilly course for competitors
> - is scenic for competitors
> - provides facilities for competitors at the start/finish line
> - has a suitable transition area* approximately half-way along the route for those running the relay option.
>
> *Transition area refers to the point where runner 1 will finish and pass on to runner 2 in the relay team. It should be easily accessible and allow access for vehicles.

Study Diagram Q5A OS Map extract of Loch Eck, Diagram Q5B, Diagram Q5C, Diagram Q5D Diagram Q5E.

Referring to map evidence from the OS extract and the resources provided:

(a) (i) **Evaluate** the suitability of the route in relation to the Race Briefing.

(ii) Identify a possible site for the transition area. You must **explain** your chosen site.

b) **Discuss** the possible social, economic and environmental impacts this event may have on the local area.

20

Diagram Q5B: Proposed start and finish area, Benmore Gardens (142855)

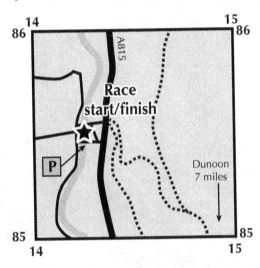

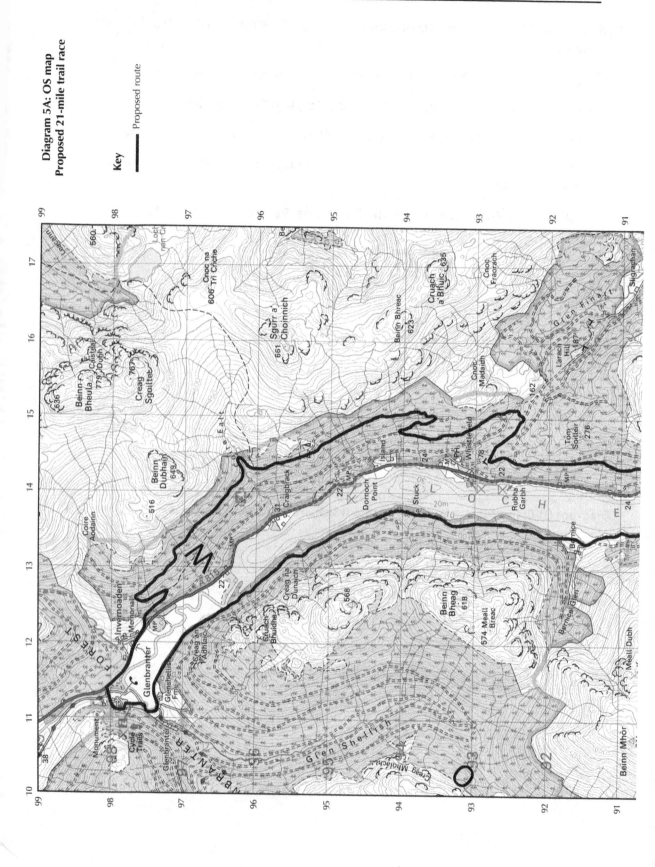

Diagram 5A: OS map
Proposed 21-mile trail race

Key
——— Proposed route

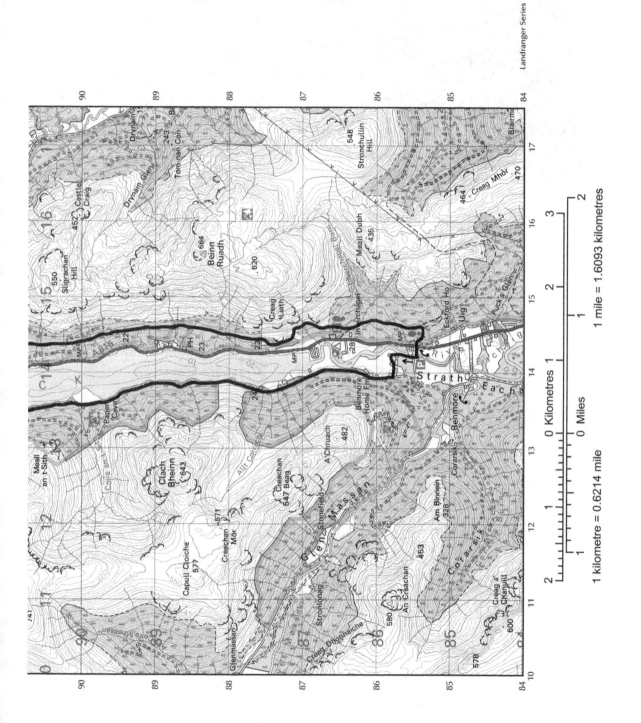

Diagram Q5D: Competitor information leaflet

Dunoon Hill Runners
21-mile Trail Race

- 21-mile Trail Race – Sunday 31 January
- Limited to 100 places in the first year
- Stunning loch-side course
- First prize male and female £100
- Medals for all finishers

For further information and race entry visit:
www.dunoonhillrunners.org.uk

Diagram Q5E: Hotel occupancy rate by month, Scotland

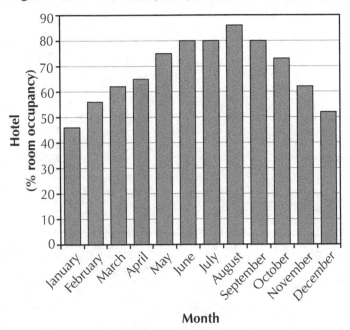

Month

[END OF QUESTION PAPER]